HOW TO SUBCONTRACT YOUR HOUSE:
Building/Remodeling

This book is dedicated to my wife. Marge. who had the additional faith and moral support I needed while going through the various stages of building our dream home.

And our subcontractors: they took a novice home builder. helped him over the rough spots. and built one heck of a house.

No. 744
$7.95

HOW TO SUBCONTRACT YOUR HOUSE: Building/Remodeling
By Lewis Blustin

TAB BOOKS
Blue Ridge Summit, Pa. 17214

Foreword

This book was written with the intention of shining light on the oftentimes confusing process of building a home. With this one book, you will have the basic working knowledge needed to build any structure—a home from scratch, a precut home, an apartment building, or even a skyscraper. From a garage to a mansion, the principles of construction are the same.

The information gathered in this book will show you how to plan your project, how to select subcontractors and supervise their work, and how to deal with financial institutions to get the money to support your project. There is also an easy-reference dictionary written just for you—the layman—so you can talk with professionals in their own language.

Most inportant of all, by applying the principles of subcontracting labor and purchasing materials as I have presented them, you can build the home you want on a budget you can afford—saving as much as 50% without compromising quality!

In short, this book provides all the resources you will need, whether you are planning to build a new home or remodel an older one. It will be up to you to decide what percentage of the responsibility you are willing to assume.

My wife and I did it, and so can you. If you need to reinforce your courage, examine the *Pictorial Review of Construction* (Chapter 14). As the old saying goes, seeing is believing.

By Lewis Blustin

Contents

Introduction

This book is not planned to frighten anyone out of building his own dream house. It is designed to show that it can be done and that, at the end of this delightful experience, you can come to the realization that it was a great deal of work, a hell of a lot of fun, and one of the most satisfying experiences of your life.

Since building a home is a very personal experience, the rest of this book will be written as if we were chatting face to face. Hence, I will be talking to you and only you.

We must begin with the understanding that in building a house from scratch there is very little mystery. It is a simple matter of putting together all the pieces of a jigsaw puzzle by finding the correct subcontractors and coordinating their areas of responsibility. You must realize that if it were impossible to do this, there wouldn't be any contractors in business today. There are many contractors going out of business each year, but I am convinced that they don't pay attention to the changing trends in their industry. The better contractors will always be building for those who can afford them.

If you are interested in a new home, finishing your basement, or enlarging your present home, and don't know where to turn, then this step-by-step homebuilding formula is for you.

Since your combination of ideas in siding, heating, electrical wiring, etc. are different from mine, I have included several possibilities you may want to discuss with each subcontractor. In contacting the various subcontractors, we discussed the information that is in this book so that we could evaluate what we wanted in our finished product. Therefore, this book is a composite of all the information we learned plus the complete factual breakdown of our dream home.

1
Background

When my wife and I first started looking for larger, older homes in the community where we once lived we realized that they were out of our price range. In addition to the high cost of these homes, the repairs that would be required in just a few years would make them too costly, in the long run, to keep up.

It was at this point that we decided to look around to see what was available in new construction. We limited ourselves to a fairly new and established community adjoining the city we were living in at that time. Our family, friends, religious ties, and my job were located in the city and we wanted to stay near to all of them.

We looked at and priced all of the new construction that was going on and honestly couldn't believe our eyes at the asking prices of the homes and the lots they were being built on. Most of them were in the high $50,000 range and more. The houses seemed to be of good quality construction and fairly nice in size, but it seemed as though too much profit was being built into them. The contractors were willing to vary the design of their houses a little in order to suit our tastes, but this was still not enough to give us the house we wanted.

Determined to get the home we desired, my wife and I started to look through the classified section of our newspaper. While reading the listings for old and new homes, we began to notice the ads placed by precut-home builders. They sparked our interest and started us thinking about custom building a home, one that would better reflect our own personalities.

Fortunately, my wife and I located a piece of real estate in the particular area where we wanted to live. So in our case the problem of finding the right piece of land was solved before we ever got involved in actually planning the home.

Our first step in contacting the precut-home builders was to request by telephone and mail their custom home brochures showing sample diagrams of homes they could build. modify. or redesign with little difficulty.

We sat up hour after hour looking over dozens of books and brochures. arranging and rearranging floor plans and exterior designs. and watching our dream house take shape. Several of the books were very informative as to the extras and options available from these companies. These included kitchen cabinets. heating/cooling systems. bathroom vanities. fixtures. etc.. as well as various styles of interiors and exteriors that we could mix and match at our pleasure.

After determining what we wanted. we mustered up enough courage to invite salesmen representing two precut-home building companies to our old home to review our plans. With the experience they had in their field. they honed the rough edges of our plans. gave us an idea of what their portion of building the home would cost. and advised us as to what we could expect in contacting subcontractors for the remaining work.

We made the decision to go with a precut-home building company and had them design our blueprints. We were very pleased with the results and planned to continue with the company. But as we approached other subcontractors to do the remaining work. they showed us how they could complete the phases the precut-home builder would cover and save us money in the process.

In our particular case we found that if we continued to go with the precut-home builder and our selected subcontractors. we would overshoot our budget by a considerable amount. In essence. the subcontractors showed us how to save money building the home from scratch by supervising our own construction and choosing and purchasing our own materials.

THE LAND AND THE HOME

What comes first. the home or the land? Actually. the land and the home should be considered together. After you make an approximate determination of the style of home you want. check the area you would like to move into to see if there is available real estate to support that style. Remember, with today's modern earthmoving equipment. property may in many instances be changed to suit your needs and preferences. Also consider whether or not your choice of style will fit in with other homes in the neighborhood.

Once you have selected the property. or think that you have. check the specific building ordinances with the building inspector's office for your area covering frontal setback (how far back the house must be set from the street) as well as side- and rear-yard setback (the amount of footage that must be left between adjoining properties). For example, check the plot plan in the home specification booklet in the chapter on financing. In order for everything to be legal the home must comply with the local laws.

When you have coordinated the land with the proposed dimensions of the home, you are ready to start the long and sometimes tedious "labor of love."

Purchasing the property can always be done quickly and easily by giving the sellers cash. If you can't pay cash, and not many of us can, check with the local banks and lending institutions to see which will lend you the money using the property as collateral. Try to get a minimum of six months time on the loan plus an extension or renewal clause written into it. This period of time should cover any extenuating circumstances that may arise.

If you are buying the property through a real estate broker, chances are very good that the institution he normally does business with will lend you the money, even if you have never done business with them before. It's the old story of who you know and not what you know. Be sure when you are finalizing the loan and having the property transferred that you have a real estate lawyer with you. This is important in having peace of mind and knowing that the property and your liability are safe and secure in every legal aspect.

Once the property is bought you can start developing the finished design of your home. Take an adequate amount of time to develop your home. Again, get magazines, booklets, brochures, and anything else you can find to help you determine the design of your home. Just remember that almost anything or any design can be built. It will cost you nothing to dream and move walls around in your head or on paper. Once you set them in wood, brick, or cement, they are permanent. So make the changes early while you still can.

When you have developed your plans, don't start building until you have all of the bugs out of them. Be prepared to live with the final plans. It will cost you much more in the long run to make changes after the building is completed. There are many architects who will be more than happy to freelance a set of blueprints for you at a very reasonable fee compared to what a large architectural firm might charge.

Unless you are on an unlimited budget, be practical with the design of your home with respect to the location and routing of plumbing, electrical, and heating/cooling equipment. Just remember that the fewer fixtures that have to be broken through to the basement walls, the more use of space you will have in your basement area. It will also be easier to section it off for additional improvements or alterations such as a workshop, laundry room, amusement room, or family room.

LICENSES AND PERMITS

Before you start building, be sure to check what licenses and permits will be needed before and during construction. This is very important to the smooth flow of your construction project. But licenses are not the problem you might think they are. There are not that many required to build a home. Simply make sure that all your subcontractors have their individually required licenses or permits to work in the community where you are building. If they don't have them they can, in most instances, apply for them at the local city, town, or village hall. There may be a nominal fee for each license and permit.

The following licenses and permits should take care of most of your needs. A brief description of each tells what is required and what it allows the contractor and builder to do.

- **Contractor's License.** This is a general license that can be applied for at the local town, village, or city hall for a set fee.
- **Building Permit.** This permit gives you the right to build your home on a specified piece of property and registers the home with the local governing body. The permit usually carries a time limit, and the construction must be started before the permit expires.
- **Water Permit.** If you are going to build in a suburban area of a large city where both areas use the same water system, you should check with the city and the suburban community to see if one or two permits are required.
- **Gas Hookup License.** Some communities require a license before you connect gas lines to your home. Generally, the heating/cooling subcontractor will notify the local gas company of your intentions to tap

into their lines. When selecting this subcontractor, check to see what his procedure is regarding the license. He may include the fee and registration in his bid, or you may have to register with the gas company yourself.

- **Electrical Hookup License.** The electric company may or may not require a license or permit, but they must be contacted a few days in advance of the construction date. They will frequently install a temporary pole on the construction site for carpenters and other subcontractors who need power to operate their tools and equipment.

As you can see, coordinating the licenses should be one of your least worries. In fact, your subcontractors will carry a great deal of this responsibility. Just be sure to check that all licenses, permits, and fees have been taken care of before construction begins. This precaution will save you time and money in the long run. An overlooked or forgotten license or permit may cause a delay in construction until the license or permit is obtained. Delays can be expensive if you have to pay subcontractors for time when little or nothing can be done. There may also be a fine or penalty for an attempt to do work without the necessary license or permit.

Inspections are also an area that should be given attention. Don't cover up or build over any area that must be inspected until after the inspection has been completed and certified. This is especially important at the rough-in stages. A strict inspection code may cause you to tear down or dismantle finished phases of construction so the inspection can be accomplished. This may also cost you time and money.

INSPECTIONS

When submitting your blueprints and home specification forms to the local building inspector, or the city, town, or village hall, it is best to get the names of any additional inspectors that may be required to inspect the rough-in and finished work.

Once you find out who the inspectors are, you should notify them to tell them who you are and what you are building. Ask if there is any additional information that their office might require concerning your building plans.

Below is a sample list of inspections that will be required in the building of your home.

- After excavation.
- Before backfill.
- After rough frameup.
- After plumbing rough-in.

- Before covering the sewer system or septic tank.
- After plumbing completion.
- After completion of the home.

Other designated inspectors, or the building inspector himself, will have to inspect the rough-in and finished stages of the electrical wiring and the heating/cooling system. Your subcontractors will tell you when their jobs are ready for inspection. It is important to have the inspection completed as soon as possible so that the subcontractor can continue with his work.

It is your responsibility to call the building inspector for his verification of each step. Remember, this is for the protection of the homeowner and insures that the quality of work meets the requirements of the local code. The tougher the inspector is, the better your home will be built.

Planning the home takes about as much time and consideration as the actual construction. A home is probably the biggest investment you'll ever make, so make that investment carefully. Most likely, you will be living in your new home for several years to come. This means anything you put into the design that may cause you regrets later should be spotted before you start building. Once the home is built you'll have to live with what you have or spend a lot of money to remodel.

Poor planning can wind up costing you more time and money than you might expect. Measurements and specifications on your plans and blueprints should be as exact as possible. That is why you should have an architect draw your plans and check your calculations. Overlooked or forgotten permits or inspections can cost you money both in fines and extra man-hours for labor. It will also cost you time since your subcontractors might have to be rescheduled if officials decide to halt your project until the matter is resolved.

So take your time. Choose the neighborhood and the style of your home carefully. And don't forget the legal requirements. You'll have plenty of time to enjoy your dream house, so don't spoil your dream by going too fast.

2

Subcontracting

How many subcontractors are necessary? Depending on your plans and specifications, your list will vary. The first step in checking this is to visit the lending institution you intend to do business with and ask for all of the forms necessary to complete a construction loan. Basically, you are only interested in the *sworn construction statement* at this time, for this form gives you a complete breakdown of the subcontractors usually needed to build a home. (See the chapter on financing for a copy of the sworn construction statement used in the construction of our home.)

Two other information forms will be given to you. One is the *home specification booklet* and the other is a *personal financial statement*. These are discussed in more detail in the chapter on financing.

Once your blueprints are finalized you can begin to line up your subcontractors. As difficult as this may sound it's just back to piecing the puzzle together. Below is a modified version of subcontracting jobs by category.

- Excavating.
- Brick and concrete.
- Plumbing and sewer hookup.
- Heating/cooling.
- Electricity.
- Carpet and linoleum.
- Plaster and dry wall.
- Stucco (exterior).
- Ceramic tile.
- Painting and staining.
- Asphalt and sod.
- Carpentry.
- Roofing.
- Exterior siding.
- Ceilings.

If you have absolutely no idea where to locate subcontractors, look in the classified section of your telephone directory, the classified section of your newspaper, and the trade association and union newspapers. The public library is an excellent place to find current editions of these publications. Combining quality workmanship with a fair price is necessary in finding the best subcontractor. You want to get your money's worth, so determine the quality of work each subcontractor does. Ask for references. Always remember: It's your money and you are entitled to all the facts.

When comparing bids, make sure you compare the job to be done and the materials to be used before selecting the subcontractor. Remember, a lower bid might mean poorer workmanship or inferior materials. Check bids carefully. Make sure that they state exactly what is to be done, what materials are to be used, and how much each portion of the job will cost. Compare your bids on the basis of all these factors and not on the merits of cost alone. Once again, you are going to be living in your house for many years to come so make your selections carefully.

Generally speaking, the more helpful a potential subcontractor is, the better you can rely on him to do a good job. When you question subcontractors don't be ashamed to admit you are a novice. An experienced craftsman can usually tell a novice as soon as he walks in the door, so you won't really be fooling anyone but yourself.

Check out as many subcontractors for each category as you think necessary. Depending on where you are going to build, it may be advisable to get a subcontractor with many specialties, such as a carpenter that can rough-in, shingle a roof, hang plasterboard, install siding, etc. Keep the number of subcontractors to an absolute minimum. The fewer workers you have to coordinate, the easier your job will be.

Most contracting companies specialize in one field, such as plumbing, heating, or electricity, but you will find some companies that combine two or more technical fields. Don't be afraid to accept bids from these larger companies. If their bid is acceptable, and you hire them, they may help you in managing your subcontractors. A larger company has more time, money, and men invested in your house than does a small subcontractor, so it will be to its benefit as well as yours to get the job done quickly and economically.

Continually check with your subcontractors for recommendations of subcontractors in other fields. This will broaden your scope and give you a lead to start off your conversation with a potential subcontractor.

At any rate, make your selection of a subcontractor based on personal experience with him, recommendations from previous customers and other subcontractors, and his bid. And remember what was said earlier about bids.

MATERIALS AND DISTRIBUTORS

Locating good subcontractors and finding material distributors are actually interwoven. One hand can, and often does, wash the other, if you don't mind an old cliche.

Assuming that your blueprints are finalized, go to an inexpensive printing firm and have at least a dozen sets of the blueprints made. These will be passed around to the various subcontractors and suppliers. The more sets you have, the easier this can be done.

Each subcontractor should have his own set of plans to work with so he can do his job and coordinate his work with that of other subcontractors. An electrician, for example, will want to know how the walls are going to run in order to plan the wiring layout; your heating/cooling subcontractor will want to know how to design the ductwork throughout the house. Each subcontractor will want to see your plans in order to visualize the completed house and design his work accordingly. This will also help them in coordinating their bids with the amount of man-hours and materials to give you the most reasonable bid possible.

Material suppliers should also see your blueprints. They are usually up to date on construction codes and customs and will be able to supply you with the proper materials for each phase of the house. They may also be able to advise you on new products and techniques that you may not be aware of and give you ideas on how to scale down your material prices with substitutes or competitive products.

MATERIAL DISCOUNTS

This step is much easier than locating good subcontractors. There are wholesalers and retailers in every field that will be more than willing to deal with you if you are willing to take the time and trouble to look for them.

Before you start your visits to these dealers and distributors, you should understand that a general contractor asks for and gets discounts on the variety of materials used in construction. As your own contractor you too are entitled to discounts. You should search for the highest quality material in combination with the biggest discount.

Listed below are samples of the discounts we received in building our house. You may not get exactly the same deals but you will have a general idea of what to expect.

- Kitchen cabinets—40%
- Lighting fixtures—40%
- Millwork—33⅓%
- Exterior siding—30%
- Steel beams—25%
- Carpeting—20%
- Garage door—20%

Don't be afraid to ask for your material discounts if they are not offered to you. After all, you are the contractor and are therefore entitled to a contractor's discount. A material supplier would rather give you the discount than not make the deal because he knows you will be able to get the discount someplace else. If a material supplier refuses to give you a discount, then shop around until you find one that will. You should be shopping around anyway and comparing bids just as you would with a subcontractor. When comparing materials, check the quality of the material, the warranties and guarantees, the promised delivery date, and the reputation of the dealer in honoring all of these factors. A big discount rate on poor materials or late deliveries is no bargain. And it might cost you more in the long run!

SURVEYS

Although the survey is a relatively simple matter, a professional is necessary. If you are entering an area that has already been surveyed, the marking stakes should be at the corners of your lot. When you check with the lending institution that is financing your construction you should also check to see if a new survey is needed to verify the original one. More than likely you will have to have a new survey taken. Besides, the survey will be needed to plot the position of your home on the lot for the approval of the lending institution and building-code inspectors.

If you are going into an unsurveyed area, you will definitely need an official survey. In any case, choose a surveyor with a well-known reputation in the area where you are building. He should be familiar with the land and be able to help you with such details as the setback, alignment, and angle of your home so it will conform to the building code and the curve of the street.

A copy of the official survey should be attached to the home specification booklet. This booklet will be submitted to the lending institution you are dealing with, along with the blueprints, the sworn construction statement, and the personal financial statement. The lending officials will use all of this information to determine whether or not you will qualify for the loan. Another copy of the booklet, the blueprints, and the survey must be submitted to the building inspector so he can be sure that your construction will comply with the building codes and regulations.

3
Carpenters and Lumber

To determine one of your major material costs, start by contacting several lumberyards. Select lumberyards that include cut price, self-service, total service, and top quality. Get a complete material breakdown for the entire construction project. This should include all lumber, nails, millwork, roofing, siding, and even kitchen cabinets if they sell them. List everything you can because it will be helpful when you go to independent wholesalers or jobbers for their bids on similar products or services.

Special emphasis should be placed on checking independent millwork manufacturers and wholesalers. Remember that they handle interior and exterior doors, windows, closet doors, patio doors, door and window accessories, and wall and floor trim. This can amount to as much as one-third of your total lumber bid. Some jobbers will give you a contractor's discount; the lumberyards may not. Again, get as many bids as time will allow.

Get bids from at least four lumberyards. Always take into consideration the quality of material, the reputation of the company in the community, and of course the bid. You are much better off spending a few hundred dollars more to get fast, sure delivery and top quality lumber. And the carpenters will be sure to have the material they need when they need it.

Lumberyards are a natural place to check for information on good carpenters interested in subcontracting their skills. Select your carpenter based on recommendations, projects of his that you have visited, and price. Ask the subcontractors to break down their bids into two categories: *rough-in* and *finished* work.

The rough-in stage should include framing the house, sheathing the exterior, shoring up the basement walls for backfill, shingling the roof, installing windows and doors, installing the exterior siding, and installing the garage door if a garage is built.

The finishing stage should include the finished flooring, casing the windows and doors, hanging and finishing plasterboard, spraying the ceilings, installing cabinets and vanities, and building a firewall in the garage.

Each component of the subcontractor's work should be on his bid. Don't take anything for granted. If something that you want done isn't listed, see that it is added to the contract. Most subcontractors are honest, but it is always better to put everything in writing to avoid any confusion or misunderstanding later.

MILLWORK

When buying your millwork, select the very best your budget will allow. The millwork includes the windows, interior and exterior doors, and all trims and moldings. It is difficult to say that one product or manufacturer is better than another. Therefore, the type of millwork you buy should depend on how much you can afford to spend, the styles and types of materials you're looking for, and where you can make the best deal.

There is an almost unlimited variety of exterior doors (see the Layman's Dictionary) that are either solid-core, hollow-core, or steel. The choice of doors is a matter of personal taste and compatibility with the design of the home. Exterior doors should be solid-core or insulated-metal doors to provide a maximum of insulation and protection. Interior doors may be hollow-core to maintain a minimum cost. While solid doors provide a better shield against the elements, they are more expensive. Hollow-core doors are less expensive and still provide the necessary privacy between rooms inside the home.

We chose a combination set of storm screens and solid-core doors, as opposed to steel doors for the exterior. We did so because our children are young and my wife wanted to leave the doors open during the warm weather to keep an eye on the kids. When you install steel doors, there are no storm and screen door units designed to fit them. The cost of a steel door as opposed to a combination set of doors is just about the same. You can use a hollow-core door as an exterior door but it isn't recommended, especially if you live in a cold or wet climate.

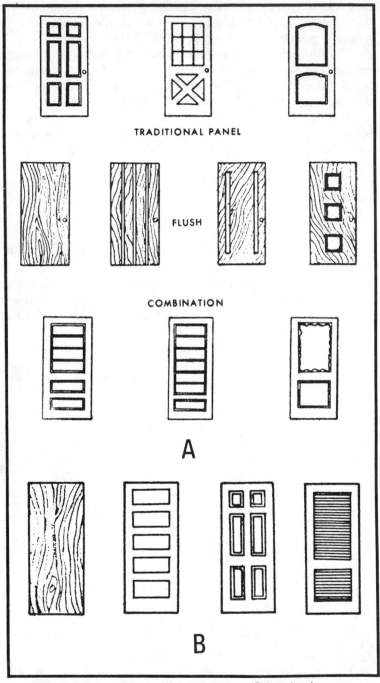

TRADITIONAL PANEL

FLUSH

COMBINATION

A

B

Fig. 3-1. Types of doors: (A) exterior doors, (B) interior doors.

If you are building an attached garage I would recommend that you install a steel door in the garage entrance to the house. In most areas, especially those where the winters are rough, VA and FHA requirements and most local ordinances will call for a metal fire-resistant door between the house and the attached garage. During the winter a lot of condensation builds up on the garage and tends to warp the wood of a solid- or hollow-core door. This will allow cold air to enter the house through the doorway. A steel door will resist warping and provide longer lasting insulation.

Because they are more economical you should use hollow-core doors inside your house, that is, unless you want a soundproof room. Hollow-core doors are the most reasonable way of enclosing rooms within your house. When you purchase these doors it is a good idea to buy them prehung even though it might cost you a few dollars more per unit. Tell your carpenter that you are buying prehung door units because his bid will probably include hanging doors. Chances are that his charge for hanging doors on the job will work out to about the same as the extra charge for prehung doors. So, even though the cost evens out, you will save time because the carpenter will be able to finish his work that much sooner. And time is money.

Regardless of what type of wooden doors you purchase, you should be sure that they are made of oak or birch and not pine or mahogany. Oak and birch will accept a better stain and stand up against scratches much better than pine or mahogony. And doors of oak or birch will look a lot richer and last much longer. (See Fig. 3-1 for different types of doors.)

Your window units will depend on the style of home you want. It is hard to tell what any one person would want because there are so many varieties. While the basic double hung window is still popular, the crank-type window and the large insulated-glass window is increasing in demand as more people are looking for individual style. (See Fig. 3-2 for different types of windows.)

If you decide to purchase double hung windows and you have a choice between a vinyl or an aluminum balance channel, select the vinyl. The windows use a spring unit contained in the channel to move up and down. The vinyl doesn't transfer cold as fast as the aluminum channel so there will be less heat lost in the winter months.

Also, you will usually have a choice between aluminum or wood-framed combination storm windows and screens. As you

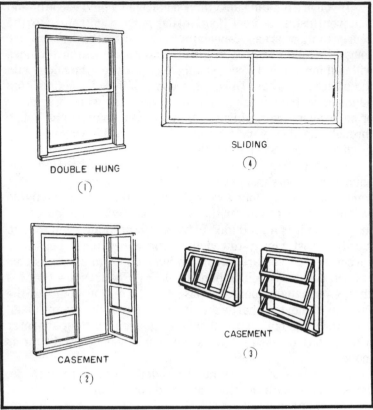

Fig. 3-2. Types of windows.

surely know, the wood units will require painting and maintenance, whereas the aluminum combinations are practically maintenance free.

Window and door accessories include window lifts, window and door locks, doorknobs and stoppers, and closet doorknobs and locks. The rest of the millwork includes baseboard trim, lattice, closet poles, and handrailings for the stairways.

Why go to a millwork company? Because you may be able to get a contractor's discount there. A lumberyard or other dealer may not be willing to give you as good a break in price. Judge the quality of the millwork you buy. If you feel you can get the years of wear out of it that you are looking for, and at a discount, then buy it. I am sure that as you learn about buying material for each area of your home, your judgment will sharpen and you will be able to determine which deals are best.

KITCHEN CABINETS AND BATHROOM VANITIES

Since you're building a custom house why shouldn't you have custom-built kitchen cabinets and bathroom vanities? There is no reason at all except the cost. From our own experience we found that some companies wanted up to $500 more for custom-made cabinets. They promised that the cabinets and vanities would blend together with the house beautifully and would be measured from the blueprints and doublechecked when the home was in the rough-in stage. All of the cabinetmakers we talked to were professional craftsmen looking to do a professional job at a professional price. The quality of the workmanship we saw was exceptional.

Somehow, though, it seemed to us that the price was just too much for the results, especially since I asked the lumberyards what the cost for the material needed to build the cabinets and vanities would be. The total cost for materials came to only 25—30% of the total bid for these items.

It was at this point that we decided to look into prebuilt cabinets and vanities. We discovered that we could get the quality and appearance we desired at a much more reasonable price if we purchased prebuilt units instead of custom-built units. Not all of the prebuilt units were of the quality and style we wanted, but we kept looking until we were satisfied. As with everything else, if you look hard enough you'll find what you're looking for.

Upon close inspection and evaluation of the cabinets and vanities we saw, I determined that the only things separating the various brands were four components: the wood, the construction, the stain, and the style. Take away the style and the stain and all you have left is the wood and the construction.

Prebuilt factory cabinets are built on the same principles as custom-built cabinets. The workmanship may or may not be as precise, but the cabinets will be built with the type of wood you specify. As in choosing the wood for doors, you would be better off selecting oak or birch instead of pine or mahogony for your cabinets and vanities. Both factories and custom shops work with charts showing the various kinds of cabinets and vanities that are made. They will both mix and match these units according to your needs and preferences. For a breakdown on the different units, check the Layman's Dictionary.

Now let's get down to the staining. It is amazing what flaws a good staining job can hide. If you are hiring a professional stainer you should let him stain *all* woodwork in your house in order to keep an even and consistent color. Your bids from the cabinetmaker or factory should exclude any finish staining because they will not be able to match the other wood throughout the house.

Once the frames of the cabinets are installed and stained I defy anyone to say at first glance if they are custom- or factory-built units. From this point it is the hardware, the design of the doors, and the staining that attracts the eye of the individual.

The style of door you choose for your cabinets and vanities will have much to do with determining the overall appearance. All factories, like custom-built shops, offer you a variety of door designs at a variety of prices. Actually, the cost of making one door over another is very small. From what we could determine , custom shops charge a little more for their products simply because of the smaller quantities they turn out. The cost of the hardware for these units seems to work on the same principle.

In addition to the standard wood cabinets, some dealers will offer a wood-grain covering over a wood base. These are usually much cheaper but it is up to the individual to choose. I prefer the basic wood cabinet because I think it gives the kitchen a warmer feeling.

It is up to you to take the opportunities to save. In addition, because you are acting as your own contractor, some factories and custom shops will offer you a discount just to get the business.

If you concentrate on getting the best cabinets for the price you can most easily afford, most people will assume that you paid a lot for them. You will have the satisfaction of knowing that your "custom-built" cabinets and vanities, be they made in a small shop or a large factory, didn't cost you the small fortune everyone thought they would.

COUNTERTOPS

Any kitchen-cabinet builder will also want to sell you countertops. There are two standard countertops that you should consider using to cover your base cabinets. One style is the S-curve countertop. This top is half as thin as regular Formica because it has to be bent and molded to fit a preformed particleboard base. It has a one-piece rounded backsplash and rounded edges. It is ideal if you have children small enough to bump their heads on the counter because there are no sharp edges or corners.

Regular Formica is too brittle to bend. It must lie flat on the particleboard base. The edges are squared and the corners are sharp. A separate backsplash can be put on the wall behind the countertop with very little effort. Each type is good and it is up to you to decide which will best fit your kitchen style.

The Formica countertops are glued to a particleboard base that is designed, either by the custom shop or the factory, to fit the base cabinet you buy. You usually get the countertops from the same dealer that sells you your cabinets and vanities. Sometimes, however, you can save money by buying these items separately. The classified section of your telephone directory has a listing for *countertops*.

The only other type of countertop you might consider using, and this only sparingly, is a chopping block. This can be special-ordered or purchased as a stock item. It is very nice to work on. We selected a chopping block to drop our Char-Glo-Grill unit into. It is much better and more convenient to trim and prepare meat in the same area it is cooked rather than carry it across the room. The cost of a chopping block countertop is comparable to that of Formica.

As for tops for your bathroom vanities, there are a number of ways to cover them. You can select a cultured-marble top, a laminated plastic top, or a custom-built ceramic top. All of these come with or without a bowl. They are defined in more detail in the Layman's Dictionary.

The cabinets, vanities, and countertops are part of the individuality that can set your home apart from others. If you take advantage of what we discovered, your home can be distinctive and elegant while, much to your amazement, staying within your price range.

Figure 3-3A is a sample layout of our kitchen cabinets and Fig. 3-3B, of our bathroom vanities. Figure 3-4 provides the sample code numbers for these units. Figure 3-5 is a picture of our finished kitchen with the cabinets and countertops installed.

STEEL BEAMS

In almost all home construction today, steel beams and adjustable columns are placed in strategic locations throughout the basement to support the structure. The number of beams and columns used will depend almost entirely on the

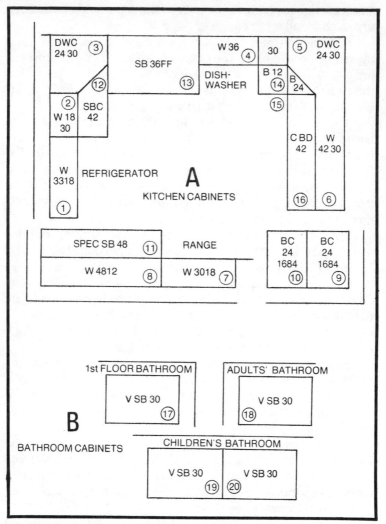

Fig. 3-3. Sample layout of (A) kitchen cabinets, and (B) bathroom vanities.

size of the home and the location of load-bearing walls that will not be based on the concrete foundation. A load-bearing wall has pressure being forced upon it from above, such as the weight of the roof or a second-floor wall.

The horizontal beams in the basement set into and across the exterior concrete block walls and act as a support while the vertical columns act as the legs. The cement men must pour footings for these steel columns at the same time they pour footings for the exterior brick walls.

KITCHEN CABINETS

(1)	Wall Cabinet	33"W × 18"L Double Doors
(2)	Wall Cabinet	18"W × 30"L Left Hand Door
(3)	Diagonal Wall Cabinet	24"W × 30"L Right Hand Door
(4)	Wall Cabinet	36"W × 30"L Double Doors
(5)	Diagonal Wall Cabinet	24"W × 30"L Right Hand Door
(6)	Wall Cabinet	42"W × 30"L Double Doors
(7)	Wall Cabinet	30"W × 18"L Double Doors
(8)	Wall Cabinet	48"W × 12"L Double Doors
(9)	Base Can Goods Cabinet	84"L × 16"W × 24"D Right Hand Door
(10)	Base Can Goods Cabinet	84"L × 16"W × 24"D Left Hand Door
(11)	Range Base Cabinet	48"W × 24"D With Two Drawers
(12)	Corner Base Cabinet	42"W × 24"D Pullout Drawer, Cutting Board, and Right Hand Door
(13)	Sink Base Cabinet	36"W × 24"D Double Doors
(14)	Base Cabinet	12"W × 24"D Pullout Drawer and Right Hand Door
(15)	Base Cabinet	24"W × 24"D Pullout Drawer and Right Hand Door
(16)	Base Cabinet	42"W × 24"D 5 Pullout Drawers, Cutting Board, and Right Hand Door

BATHROOM CABINETS

(17 – 20)	Vanity Base Cabinets	30"W × 21"D Double Doors

Codes: W = Width L = Length D = Depth
All Base Cabinets are a standard height of 34".
All Base Cabinets must be blended into a design in conjunction with the countertop.

Fig. 3-4. Sample code numbers for cabinets (Fig. 3-3A) and vanities (Fig. 3-3B).

Fig. 3-5. View of finished kitchen with installed cabinets, countertops, and appliances.

The steel beams and columns are the first items the carpenters install before they cap the foundation. The beams are placed in a permanent position inside predetermined, partially-cut cement blocks. These are cut by the cement men according to blueprint specifications after they build the walls. Although the columns are set in place they are not made permanent until the concrete floor is poured. This is done because it might be necessary to move them an inch or so either way in order to maintain maximum coverage of the load it must bear.

The beams and adjustable columns are easy to purchase. Most urban and suburban areas have steel suppliers that sell them. If you are stumped, ask your lumber dealer. At one time or another he has had a use for them and can usually refer you to a company that sells them.

How do you determine what to buy? If an architect designed your plans he should have sufficient engineering knowledge to advise you on what size and weight is needed. If not, all good steel suppliers have engineers who can calculate from your blueprints how much stress will be put on each beam, and they will sell you beams that will do the proper job.

There are also adjustable columns. They are adjustable because basement depths may vary when the hole is dug and the footings are set. There are various sizes of columns that can be adjusted about three to four inches. Again, a good consulting engineer with the steel company can tell you exactly what you need if he reads your blueprints. To be doubly sure, it never hurts to check with more than one company and take bids in this area just as you would with any other.

4

Bricklayers and Cement

A good foundation is very important to the house. Also, it may eventually support an additional living area. Therefore, though you might be able to hedge with some subcontractors, be especially careful to get the best bricklayer (mason) you can find to build the foundation. The carpenters you deal with will know many brick and concrete men and should give you the names of a few to contact.

Get your brick men to give you individual bids on constructing the basement walls, cement floors for the basement and garage, exterior facing, fireplace, and stack. Make sure these bids include footings for the basement walls, steel columns, and basement windows.

When considering the basement plans, you should be aware that footings are the base of the cement block wall. They are the first items formed and poured (Fig. 4-1). They should also contain steel reinforcing rods (rerods) for added strength to hold the footings together and resist cracking under pressure (Fig. 4-2). All cement will eventually crack and shrink to some degree. Also, cracking is partially caused by the settling of the ground under and around the home. Footings poured for steel columns don't require rerods unless they will have to support an unusually heavy load. Eventually, they are totally buried beneath the cement basement floor (Fig. 4-3).

Your garage floor should include a wire mesh (Fig. 4-4). The wire mesh serves the same function as the rerods. It holds the cracking of the cement floor to a minimum and prevents major problems in the coming years.

Ask the brick man to bid on an 11-course basement wall. A course is one row of blocks, hence an 11-course wall will mean the blocks are stacked 11 rows high.

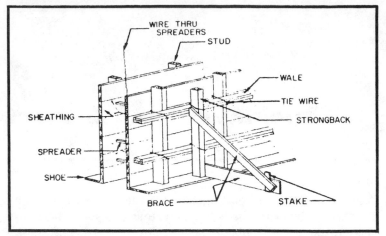

Fig. 4-1, Wooden forms are built to pour concrete to form concrete footings.

Two kinds of cement blocks, modular and standard, are used in the construction of basement walls. Modular block dimensions are 15⅜ × 11⅝ × 7⅝ in. while the standard block dimensions are 15⅜ × 11⅝ × 8 in. The full eight-inch high standard block is recommended for more head room. The 11-course modular block wall will give you a total of 7 ft 5 in. height from the base of the footings to the top of the eleventh course. Add 4⅛ in. to that figure if standard blocks are used. But after your four-inch thick concrete floor is poured you will be down to a height of 7 ft 1 in. using modular blocks and 7 ft 5⅛ in. using standard blocks. As you can now see, building a

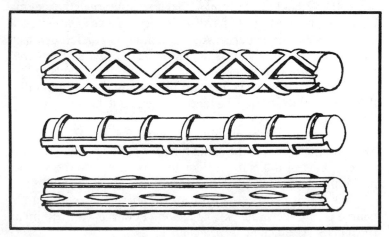

Fig. 4-2. Types of reinforcing rods used in concrete and masonry.

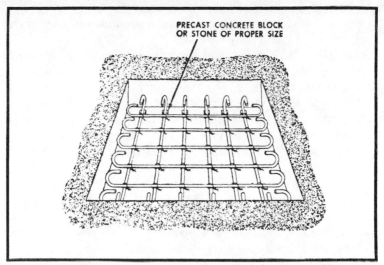

PRECAST CONCRETE BLOCK
OR STONE OF PROPER SIZE

Fig. 4-3. Steel columns used for support and placed in footings.

basement with head room any lower than this would almost eliminate any future plans for additional living space.

In addition to the four-inch deep cement floor subtracted from the total height of the wall, if you are using a forced-air ductwork system attached to the ceiling of your basement you must allow at least a nine-inch drop for when your basement heat registers are added. Building any basement with less than 11-course basement walls will end up being a poor investment and a negative feature at resale.

If your plans include a fireplace or a partial- or full-brick exterior, it is advisable to have the same subcontractors do all the work. The headaches are less and the price may be better

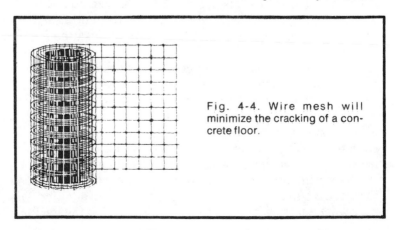

Fig. 4-4. Wire mesh will minimize the cracking of a concrete floor.

because one subcontractor would be more indebted to you and schedule you into a busy calendar if you give him more work, especially since his crew must come back at different intervals anyway (weather permitting) to complete their various assignments.

It is also advisable to check with you local brickyards for the price on concrete blocks. Compare your cost to what your subcontractors want to charge. You may find tremendous savings or they might be so small that it is not worth the trouble to order them yourself. It all depends on how much profit the subcontractors want to make on the bricks.

For finishing bricks, a mason will incorporate a charge of about $100 per thousand bricks in his bid. Actually, this is the average additional cost of a good eight-inch finishing brick. These will vary in price from $90 to $110 per thousand, with bricks 12 in. long varying from $120 to $150 per thousand. If you use the 12 in. brick exclusively, the cost compared to using the eight-inch brick will average out about the same for the area you want to cover because you naturally use fewer 12 in. bricks. One thousand eight-in. bricks, for example, will cover about 125 sq ft of wall space excluding mortar, whereas a thousand 12 in. bricks will cover about 187 sq ft.

When you are considering the finished appearance of the brickwork, remember that the mortar between the bricks can be made part of the design and doesn't have to cost extra. You can use the traditional concave, tooled appearance or a sculptured weeping, uneven look to give an unusual finish. The weeping mortar style should be used only with a white brick because when it drys it is difficult to clean. Bricks of another color combined with weeping mortar will be spotted and leave a messy appearance. The traditional concave mortar style is standard for the exterior finish but is not always used for the fireplace.

Be sure to discuss mortar colors, joint tooling, and final cleaning with your masonry subcontractors. These are important items in the total cost of their bid.

If you want a fireplace in your home (see Fig. 4-5), decide also if you want a mantle or a hearth that is either raised or flush to the floor. The price may vary from $50 to $75 for the hearth, with the mantle costing more. The price of building the fireplace will also be determined by the number of flues. Each fireplace requires a separate flue and damper. In addition, you may want a larger chimney with dummy flues at the top for a

Fig. 4-5. Brick fireplace with raised hearth and no mantle.

more massive appearance. The unused flues are strictly for show. If you have only one fireplace in your home I recommend you build it with two flues. This will give a plush appearance.

Last but not least is the sidewalk. It is difficult to decide what to do about installing it at the time the home is constructed. The problem is that the ground around your home will take some time to settle. Most of the cement men I've talked to didn't even want to include the sidewalk in their bid. In fact, most of them wanted to wait from six to nine months before installing any sidewalk on the property. They claimed that the settling of the ground would cause the cement to crack and they wouldn't guarantee any sidewalk installed during the construction of the home. I feel it is better to be safe than sorry. If you plan to have a large amount of sidewalk concrete installed, it may be advisable to wait several months after you have moved into your new home. In any case it won't be too much of a problem until you begin to landscape the yard.

If your subcontractor is hesitant to lay a concrete sidewalk on newly graded ground, ask him about using a vibrating soil compactor to settle the area that will support the sidewalk. It is important that this ground be sufficiently compacted to avoid all but minimum settlement.

The excavator can do the actual compacting job after he has finished the backfilling or grading. Be sure your cement subcontractor checks the soil compaction before he pours the sidewalk. A simple test of pushing a half-inch thick reinforcing rod into the ground will tell how firm the ground is. If he can push the rod down a few feet by putting his weight on it, the ground is too loose and will have excessive settling.

It is a good idea to discuss, at the time you get your bids, whether there is any additional cost involved in using a soil compactor should it be necessary before the sidewalk is installed.

POURED CONCRETE WALLS

It may not be the wave of the future, but some builders are testing the possibilities of using poured concrete for foundations instead of the standard concrete block. Since the principles of building this type of wall are the same used by large construction companies in building many commercial buildings, the dependability of such a wall should be without question. One subcontractor told me that engineers have rated it six to ten times stronger than a concrete block wall after the backfill is in place. The poured concrete wall should also be shored up before the excavator backfills against the basement.

When I checked into the possibility of using poured concrete walls I was told that the wall need only be 8 in. wide compared to a 12 in. wall for a concrete block foundation. In constructing the poured concrete walls you must first set your footings the same as if it were a conventional wall. Next, two parallel rows of prebuilt wooden forms are placed 8 in. apart on the footing. They are locked in place and form a giant mold in the shape of the future basement walls. Basement windows are set in the forms before the walls are poured.

Wet cement is poured directly between the forms and left to dry for about three days. Rerods are first placed in the forms at the three and five foot mark, above the footings to prevent extensive cracking. At the end of the drying period the forms are unlocked and removed. Some forms now have a design on one side to give the appearance of a concrete block wall.

Since water can work its way through anything, given enough time, have the underground outside area sprayed with hot tar or asphalt to dampproof it. For extra insurance don't forget to install the polyethylene liner over the cooled tar or asphalt.

The cost of a poured concrete wall is similar to that of a standard block wall. What you gain is added strength and durability. It will not crack as easily as a concrete block wall.

The only drawback to a poured wall is that it is hard to pound nails into. If you want to finish the basement you must use a studgun to help nail and fasten the studs to the wall. Or you can try some of the new glue products to anchor the wall studs to the concrete wall. Some subcontractors have told me that people don't like the appearance of the wall, but if you intend to finish your basement it won't make any difference.

This comes under the planning stage. Think about what you want. Decide what kind of finish will please you the most and consider the future. You may want to remodel, redesign, or improve your basement. So, plan to have it built with these and any other personal considerations that will concern you and your home.

DAMPPROOFING OR WATERPROOFING

Dampproofing is an extremely important and necessary step toward the total use of your basement in the years to come. It is also relatively inexpensive. It's a messy job, but you may be able to have it included on the bid of the cement men. If you can't get them to bid on it, you have a few alternatives.

You may have to buy a few five-gallon cans of tar and a broom and put it on yourself, or you can check the local newspapers and telephone directory under the waterproofing and dampproofing classifications to find companies that perform this service. The companies usually charge about 11 cents a square foot to spray hot tar on the exterior of the basement walls. For example, if you have a full-depth basement (at least 8 ft deep) and your wall is 40 ft long, are building a split entry, your basement depth will be 4 ft. Remember to cover only the area that will eventually be covered by backfill.

In our case, we used a company that force-sprayed hot tar on the outside of the basement wall. It was done after the mortar dried and before the dirt was backfilled against the outside walls. We chose this process because the hot tar sticks quickly and easily to a rough surface. The tar is put on with an even consistency and you can get a thicker coat. One coat would usually be enough. If it is done by hand it may take two or three coats to get it on thick enough and to make sure every spot is covered.

Even though this tar cover is standard operating procedure, it is not guaranteed to prevent water from leaking in. Sooner or later the soil and water beneath the surface will eat away at the weaker spots and cause you trouble.

You do have an alternative. Simply wrap the basement in plastic.

Buy one or two 100 ft rolls of six-mil thick polyethylene plastic. It costs around ten dollars a roll and can be purchased at the lumberyard where you buy your other materials. The roll unfolds to a depth of 8 ft. The tricky part comes when you unravel it and drape it to cover the tarred area of the basement walls. Start in a corner close to where the dirt and foundation meet in the unexcavated area (usually by the attached garage, if any). Carefully drape the end over the spikes sticking out of the top row of the cement blocks. These spikes are about four feet apart. Between each of the spikes, place a stone or large chunk of dirt for more support. It won't be so easy for the polyethylene to tear that way.

As you are hanging the polyethylene, carefully smooth it out and stretch it. Keep it as even as possible to prevent any creases or air pockets from forming between the plastic and the wall. Press it against the tar; it may stick a little to it. At the same time, at the base of the footings keep the polyethylene even and lay lots of loose dirt over the bottom to keep it flat against the wall. This process will anchor it solidly all around the walls where they were tarred.

After the roll or rolls are in place, make sure your supports are strong. You may have to wait a day or two before the excavators return to backfill the dirt against the foundation and sandwich the polyethylene in place. The combined cost is relatively small, most likely under $100.

5

Plumbers and Electricians

For the following two subcontractors, the plumber and electrician, I am going to advise you to do something different than I actually did myself. It is extremely easy to see the plumbers and electricians and get bids on labor and material the same as one should do with the heating/cooling subcontractor. But I had no idea at the time I received their bids that they could rough-in and finish a house as fast as they did.

Cost wise it may be more advantageous to hire a plumber or electrician by the hour instead of the job. Buy the necessary licenses if they are not licensed in the area where you are building. As for material, do the same as you did with the lumber company. Take your blueprints to several plumbing and electrical supply houses. They will break down the material you need for rough-in and completion. Don't be afraid to ask for your contractor's discount.

Then compare. Determine if you can save enough to make it worth your while to hire a plumber or an electrician by the hour or to accept their bid for the entire job using their materials.

Check the Layman's Dictionary under the electrician's and plumber's categories for a breakdown on what to look for in materials to be included in the suppliers' bids.

PLUMBERS AND MATERIALS

Before getting any bids from a plumber, check the local plumbing section of the building code. With new innovations in the building industry coming out every day, the codes are constantly being revised. Today, plumbers *have* to use cast iron pipe below the basement concrete floor, *want* to use plastic pipe above the basement floor level for waste and water return to the sewer, and *have* to use copper tubing to bring water from the main supply into the house.

Many plumbers today will bid for jobs only if they can work with plastic wastepipe above the basement floor level. Plastic pipe is advantageous for the plumber to use because it is lightweight and saves precious time and energy. The smooth interior of plastic pipe doesn't catch any waste particles being flushed through it as does the cast iron pipe. The plastic pipe is very easy to glue, seam, and repair. The only disadvantage is that when extremely hot water is permitted to run through plastic pipe 1½ in. and smaller in diameter for any prolonged period of time, the pipe may develop a sag. With the use of thin wire braces placed in strategic locations, this problem is minimized. A good plumbing inspector will catch this during the rough-in inspection.

SEWER, WATER, AND CESSPOOL HOOKUP

If you are moving into an area that already has sewer and water lines available, chances are good that the local code will allow you to hook up immediately or within a given period of time. The most reasonable thing to do is to hook up as soon as possible. This is a fairly technical and mechanical procedure and is checked by the building inspector.

Most plumbing bids will include the sewer and water hook-up as a separate charge from the rough-in and finish work because the work is done outside the home. The plumbers will give you a flat bid, depending on the setback (how far back the front of the house sits on the property from the street).

It is advisable to locate the engineering firm that designed and installed the local sewer system (or the village or city engineer) because your plumber will have to locate the exact points where he is able to cut into the lines. The engineers will be able to direct the plumber to the connecting entrance that will enable him to coordinate the hookup.

Because gravity is involved in getting rid of the internal waste through these lines, it is important that someone with experience do the job. The pipes must be set at a specific angle and most professionals are well equipped to do it or know other plumbers that specialize in it. Your plumbing subcontractor may contract it out to a specialist, which is fine as long as it passes the inspection.

Remember, a water permit is needed before you can hook up to the local system. At this time, when you apply for your permit, the local water department should advise you of any water pressure problems in your area. If not, be sure to ask if any exist. If the pressure is over a certain point (usually 85 psi), they should advise you to add an inexpensive water-pressure reducing valve, installed along side the water meter. Otherwise, the high water pressure destroys your water pipes, tank, toilets, and sinks, and it takes a very short time for the damage to start showing.

Before you do anything about a cesspool or septic tank, or both, check with the local building or plumbing inspector for the area you are moving into. Find out what the local code will and won't allow. He will know if a percolation test, a test to find out how quickly the ground absorbs water, has to be taken on your property or if there is one on file. You may have to hire a qualified engineer to perform this test.

At this point you will know what kind of plumbing units you will need installed. Check with your local telephone, gas, and electric companies to find out if any lines or pipes are buried anywhere in the vicinity of where you are planning to set the cesspool. Don't skimp in this area. You will give the cesspool continuous use and it will only be a source of constant problems if it gets clogged. Study the various models and designs available and don't be afraid to ask questions. In addition, be sure to check the depth of the water table on your property. This could also be a source of conflict in the future.

ELECTRICIANS AND MATERIAL

Your first step should be to check the local electrical code. Again, because of new materials being developed everyday, the codes must be evaluated.

Until the past few years only one form of wiring was accepted by the code. This was called the *Greenfield* system. Now, *Romex* wiring has been okayed for use in many areas. When getting your electrical bids, get separate estimates on each system from the same electrician. The bid may vary several dollars an outlet by installing Romex rather than Greenfield wiring.

What are the Greenfield and Romex systems? Greenfield, the older one, is a system where the wiring travels through ¼ in. conduit tubing, running through holes drilled into the rough-in studs, from the switches to their outlets. If it ever becomes necessary to repair wiring in your home, you can easily make the repairs if you installed the Greenfield system of wiring. You can string new wire through the conduit tubing with great ease.

The Romex system uses wire covered with a thick plastic coating. The plastic replaces the ¼ in. conduit. The wire runs through the same holes drilled into the rough-in studs that would have held the conduit. This method saves on the cost of material and labor. With Romex wiring, however, it is next to impossible to rewire an area without tearing down the wall.

The Romex system is becoming code-approved in many areas today because it reduces labor costs for large scale contractors on industrial and commercial construction and large housing projects or units. My advice would be to stay with the Greenfield system for the relatively small difference in cost.

Electrical contractors will usually charge you a flat rate of so much per opening for each connection. The subcontractor will give you an estimate taken from your blueprints of how many openings he thinks you will need and where they must be placed according to the local code. Then it is up to you to individualize the setup. An opening is counted as a single switch, fixture, or duplex wall outlet. Double switches count as two openings, triple switches as three openings, etc. This is because more than one set of wires must be run into a multiple switchbox. The electrical subcontractor's bid will include the wiring of the following standard items. You may decide which items to add or subtract from the list.

- Service box.
- Regular openings.
- Furnace.
- Well and pressure tank.
- Electric range.
- Hood fan.
- Dishwasher.
- Garbage disposal.
- Doorbell chimes.
- Electric dryer.
- Electric hot-water tank.
- Clock receptacles.
- Bathroom vent fans.
- Tv outlets.
- Central air-conditioning.
- Electric heat.
- Light fixtures.
- Electronic air filters.

LIGHT FIXTURES

Any electrical contractor generally allows a certain dollar allotment toward the light fixtures even though he knows this figure is unrealistic. This is one area where the contractors can shave a few hundred dollars or more from the original bid on your home. They know with the retail cost of fixtures today that, unless you stick with the bottom of the line on all fixtures, there is no way to beat their figure. They are also relieved of any responsibility for adding to the cost of your home because you wanted better lighting fixtures.

What we discovered, and you must realize, is the actual cost of lighting fixtures. A contractor generally gets a 25—40% kickback from the retail lighting companies that sell fixtures. For example, the contractor allows you $400 for all of your fixtures. You eventually spend $1,000. The contractor adds $600 to the cost of your home and receives $250 to $400 from the lighting company, depending on the kickback agreement.

When you are developing your home, allow yourself a $400—$600 budget for your lighting fixtures. See your local lighting-fixture companies and don't be afraid to ask for your contractor's discount. The discount saved us $400 on our fixtures and the cost of our home. By smart shopping you can make some super buys, keep within your price range, have a greater selection of merchandise to choose from.

6

Heating and Cooling

If you are not sure what type of fuel you want to use in heating your home, I suggest you contact subcontractors who handle the various ways of heating and cooling. The four major fuels are gas, electric, oil, and LP (liquid petroleum) gas. The three main heating systems are hot water, forced air, and electric. Remember to consider what is available in the area in which you are planning to build.

HOT WATER

Of the three main heating systems hot water heat is the least used. It requires a boiler with an expansion tank hooked up to baseboard units via copper tubing. It is a fantastic way of heating your home, but it has the drawback of not being able to be hooked up with a central air-conditioning system unless a separate ductwork and cooling system is installed during the rough-in period. The cost of installing ductwork and the hot water system at the same time may be prohibitive. If you stay with the hot water system, your best bet may be to use window or wall mounted air-conditioning units. The hot water system usually has a series of two copper pipes for each zone in the house. One pipe is for delivering hot water to the rooms while the other pipe returns the water to the boiler. The copper tubing, starting from the boiler, runs along the baseboard area of each room. Aluminum plates with holes in the center are attached perpendicularly to the pipe before it is put into position. When the hot water travels through the pipe it heats up the aluminum plates which in turn heat up the room. A baseboard unit covers the pipes and aluminum plates.

The system is automatic via a thermostatic control unit connected to a circulating pump at the boiler. The pump starts when the room temperature drops, allowing hot water to circulate through the baseboard unit until a prescribed room temperature is reached.

FORCED AIR

Forced air, either gas or oil powered, is the most common heating system used for homes today because a central air-conditioning unit can be economically included by placing the cooling coil inside the furnace plenum chamber. And this cooling system can use the same air distribution ductwork that is designed for the heating unit. In addition, the air-conditioning system's condensing coil can be conveniently located outside the home. The condenser is connected to the furnace coil by insulated copper tubing.

The size of the furnace and its blower, as well as the cooling coil and condenser, will be determined by the heating subcontractor. He calculates the amount of ductwork, the number of registers, and the type of heating and cooling unit by using a standard heat-loss formula. The formula is based on the size of the individual rooms along with the number of windows in each room.

Forced air is the most economical system to use. The main disadvantage is that it doesn't provide an even flow of heat as does hot water or electric baseboard heating.

If you are going to use oil or LP gas to power the forced air furnace, be sure to leave enough money in your budget to buy or rent the additional tank required to store the fuel. Also, determine where you are going to store the tank. Again, check with the local building or heating inspector; he will give you the legal alternatives. Should you use these types of fuel, remember to consider the storage tank when developing the blueprints and your potential landscape.

When considering LP gas, the storage tank should be no closer than ten feet from your house. There are two sizes of tanks—500 and 1000 gal. units. If your furnace is over 90,000 BTU in power, a 1000 gal. tank should be installed. It is important to have the larger unit because, depending on weather conditions, there will be some condensation and evaporation that will affect the pressure flow of the LP gas. The more gas you consistently have in the tank, the closer the pressure will remain at maximum efficiency.

Since LP gas companies don't like to sell the tanks, there is an installation and rental charge for them. Usually, the installation charge includes the first year's payment. The rental charge for the 500 gal. tank is around $25 a year, while the 1000 gal. tank is around $40 a year.

It is up to the homeowner to check the amount of fuel in the tank. When it gets to a low point he must call the LP gas company for a delivery. Some companies have an automatic refill service but it is never guaranteed because of the possibility of trucks breaking down and other unexpected problems. LP gas is cheaper to use than oil if you have a gas-powered water heater and other gas appliances. If the major appliances and water heater are run by electricity then oil is cheaper to use.

Oil heat and LP gas systems are similar in many ways. But oil heating tanks can be buried in the ground, stored above ground, or placed in the basement, whichever you desire and the local code will permit.

Oil heating subcontractors generally prefer to sell storage tanks as opposed to renting them. If you plan to use oil heat and want to bury the storage tank, code approved, have your excavator dig the hole when he is excavating the foundation.

Oil storage tanks are available in 550 and 1000 gal. sizes The large tank costs around $150 while the smaller one is lower priced. The subcontractors recommended to me that the tank be filled to the ⅞-full mark during the summer to allow for expansion. Also, a buried tank will have a minimum loss of fuel because the condensation is kept to a low level.

Oil subcontractors claimed heating with oil is approximately 15% cheaper than heating with gas. I haven't any way of verifying it, but the subcontractors had figures to back up their claims. Also, they claim that the average size room will require 200 gal. of oil a year to heat and cool it. The oil companies would like to fill the tank with a delivery of 600 to 800 gal..With a load of this size they usually give some type of discount. Remember, oil heat or LP gas can be used as the energy source for not only a forced air system but any other system that can be installed in a home.

A fuel powered furnace is more costly to install than one powered by electricity. But a home heated by electricity requires more than normal insulation and this brings the cost back up. It would be advisable to check with the electric power company in the area where you are planning to build for the exact insulation specifications needed for gas and electrically fired furnaces. Also notify your lumberyard, before they give you a bid on the materials, of the energy source you are planning to use to heat your home.

Along with the actual forced-air furnace, you have an option of additional equipment to make living in your home a more pleasant experience. These include the following options:

- Electronic filter. Eliminates 90% of the dust in a house, provided the filter is cleaned every month.
- Humidifier. Keeps the air moist in the winter time; it also prevents things from drying out and eliminates static electricity.
- Air-conditioning. Gives you an even flow of cool air throughout the house when the warm weather sets in.

Any forced-air system can be adapted to include any or all of these additional units.

ELECTRIC

A third form of heating has taken on some popularity recently—electrical baseboard heating. The setup is similar to hot water heat, but a heating coil replaces the copper tubing that circulates hot water. Again, there is no ductwork necessary for this form of heating. Your local power company will give you a free estimate on installing this system.

ZONE HEATING AND ALTERNATIVES

If you plan to build a home that is larger than average you may want to consider zone heating or possibly two furnaces. Zone heating enables you to keep different parts of your home at different temperatures. It requires the use of an individual thermostat for each zone in the house. Depending on the number of zones you install, you are adding on a large additional cost. Not only is it more expensive to install, but it is more expensive to repair.

Zone heating involves individual motors that open and close dampers in the ductwork. These motors usually work fine for at least five years, then they may require repair or replacement. Furnace men have told me that most service calls on furnaces involve the zone heating units.

What is a damper? A damper is a metal flap inside the duct that adjusts to completely opened or closed to provide or prevent airflow through a duct run. The damper must be properly set in order to get maximum efficiency from the heating unit.

Instead of zone heating, check into the possibility of using two heating and cooling plants, each using a single zone. If you consider two furnaces, neither one will be near the size of one furnace with two or more zones. In fact, both may come close to equaling the size of a large furnace.

There will be short periods when both furnaces will be used; but, basically, only one furnace will be running during periods when heating and cooling is required. This can mean a savings on the heating bill.

For example, most heating men wanted to install a furnace with 180,000 to 200,000 BTU of power and three zones, one for each floor above ground and one for the basement. In addition, they wanted to install a 48,000 BTU central air-conditioning system. They felt that one system could take care of all three floors. Instead, we chose a single-zone 135,000 BTU forced-air gas-fired furnace to heat the first and second floor. Using this size furnace we were able to combine it with a 37,000 BTU central air-conditioning unit because there would be less chance of it overpowering itself and freezing up at night.

We installed a 75,000 BTU furnace to heat the basement. Thus, the basement furnace need only be lit and working during the cold months when the basement is being used. The 135,000 BTU furnace combined with the forced air unit is doing an excellent job of keeping the two upper level floors warm in the winter and cool in the summer. Remember, hot air rises and cold air falls, so part of the heat that keeps the first floor warm in the winter will also heat the upper level.

To save even more on your heating bill get bids with six-inch ductwork instead of the normal five-inch ducts. You will get 20% more heat and your furnace blower won't consume as much power.

If you are contemplating building a split-level home don't fail to install a warm-air duct in the foyer or entrance way, and have it situated so warm air is circulated near the doorway. Without this heat duct, during periods of cold weather you will have a tremendous heat loss and your lower level will be colder than you would like to have it. The warm-air duct should also be situated in the foyer of any home that has double front doors. Double doors permit more than normal amounts of cold air into the house during the winter.

The heating apparatus is very critical. I would strongly recommend getting several bids, seven or eight as a minimum, and letting the subcontractor buy and install the ductwork system. The subcontractor has to compute the heat loss in your home, the best placement for cold and hot air returns, and which walls and studs are free to accept ductwork straight up and down through all levels of a multilevel home.

It is advisable to let your heating subcontractor be the first to start working after rough-in has been completed by the carpenters. The electrician and plumber can work around the installed ductwork because they can be more flexible with their material.

If you are building a single story home your plan, as far as your heating subcontractor is concerned, is much more simplified for the ductwork set up. Make certain, regardless of what size house you have, that you install the correct number of cold-air returns. This is to insure that your home will have excellent air circulation and remain adequately heated.

If your single-story home is going to be unusually long, you may want to consider two parallel heating/cooling plants because the hot air coming out of a furnace loses its heat the further it must travel. Using this system you again have the option of varying the temperatures in different parts of your home, insuring yourself of adequate heating and cooling, and enjoying the pleasures of zone heating with none of the anticipated repair problems.

Selecting a contractor eager to explain his ideas in heating and cooling your home is a step in the right direction. Generally speaking, the subcontractor that takes the time to tell you about his system will take the time to install a system that will give you good service.

Again, the power source you select to run the furnace should depend upon what you prefer, what is available, and what is the most economical to use while getting the job done.

7

Outside the House

Starting from the roof and working down past the siding, gutters, and downspouts, we can finally discuss landscaping. After all, we would like to be proud of our dream house and the outside is the one area that will be seen by friends and strangers alike.

ROOFING MATERIALS

This could be an area where you may be able to save some money. It is important that you get your bids from roofing specialists, but also have your carpenters include the finished roof in their bid. The basic principles of roofing a house are similar to roughing-in the house. It is just a bit more tedious. If you select a good carpenter to build your home he should be more than qualified to install the roofing shingles.

To have the safest possible roof from water leakage and melting snow or ice backup use 15 lb asphalt roofing paper as a base layer over the exposed rough-in plywood roof. Next, lay 43–60 lb asphalt roofing paper over the first three feet up from the facia and all around the roof. Then lay 30–50 lb asphalt roofing paper on the next three feet up from the facia and all around the roof. This double protection covers the area where any backup will occur and where most of the weight of snow or ice will accumulate.

When buying roofing material, include a roll or starter strip that matches the color of the shingles if you use the standard 3-tab design. The starter roll goes on first, just above the facia. Since standard 3-tab shingles have narrow cuts in them, the starter is used to fit in underneath the first row of shingles; otherwise, the black color of the asphalt paper will show through. By using the starter roll the roof will have the finished look you desire. There are several different styles of roofing shingles on the market today. It is important that you get all the literature you can from lumberyards and specialists that supply roofing materials.

Determine by talking with the architect how much material weight the roofing structure can support. Then select the correct shingle and weight necessary to do the job and accent the design of the home.

In almost all cases the FHA and VA require all shingles to be nailed in place, but check with your building inspector to see if the shingles have to be nailed down or if they can be stapled. In our case the code stated that they had to be nailed down, which is my preference also. It may cost a little more to nail down the shingles, but I believe it leads to a better job and a longer lasting roof. Also, check your code to see what is the proper minimum-pound test shingle to use. Generally, a 235 lb test shingle is the minimum weight you can use. It is durable and probably your best buy. The test pound figure is the weight of the number of shingles it takes to cover any 100 sq ft of the roof.

Depending on your budget, unless the roof is an integral part of the design don't put too much emphasis on it other than buying the best material, color coordinating it to the rest of the house, and having it installed in a proper manner for a carefree life of 20 to 30 years.

SIDING AND SUBCONTRACTOR

The selection of siding for your home is a matter of personal taste. There are a number of choices you can make and it all depends on which siding fits the style of your home. The variations include aluminum, brick, rough sawn cedar, shakes, steel, stucco, vinyl clad, and wood. New exterior products and designs are being developed every day. All of them can be used by themselves or mixed and matched to get the desired exterior finish. (See Fig. 7-1)

Prices will vary. Your cheapest, in the short run, are shakes. But, in turn, shakes are probably the most costly siding to maintain because they require painting every four to six years. At the other end of the maintenance scale is aluminum and steel siding, with a guarantee of 20 to 30 years' carefree use, or brick and stucco which should be good for the life of the house. Also, rough sawn cedar and vinyl clad are excellent when it comes to semimaintenance-free living.

Rough sawn cedar is a natural wood and is subject to shrinkage and expansion depending on the weather. In cold

Fig. 7-1. Finished house with combination brick and aluminum siding.

climates it must be installed by craftsmen who care for and understand the product and allow for its movement. With the exceptions of brick and stucco, siding can be installed using the basic principles of carpentry.

Steel and aluminum sidings are usually installed by people specializing in the two products, but here again good carpenters can install it with minimum trouble. Just remember, aluminum can be cut with an electric tool because the sparks caused by the cutting do not harm the aluminum siding. Steel must be cut by hand. Sparks that land on the steel surface will fuse almost on impact or burn away part of the finish. Installation costs on steel siding may run a little higher. Also, steel itself may cost more because it is a few thousandths of an inch thicker than aluminum.

If you are interested in using brick, either for accent or for the total home, inform your brick and cement men. As for exterior stucco, if your bricklayers do not do stucco work they generally know subcontractors who specialize in that field. Before using brick on your exterior, determine through the local tax assessor what additional tax assessment may be needed per square foot of brick coverage. Some areas feel a home with a brick exterior, either full or partial, is worth more than a home sided more conventionally.

Siding distributors base the cost of material on the outside square footage of the house. Siding is usually sold in units to cover 100 or 1000 sq ft minimums. For an estimated figure, multiply the height times the length for each side of the house. Then add the dimensions for each side. Subtract the total area of the windows and doors from the total of the outside dimensions. This will give you an approximate figure to determine how many squares (a square covers 100 sq ft) of material you will need to cover the exterior.

When talking to a local siding wholesaler remind him that you are the contractor. Remember, you are buying at the same place mechanics and contractors do. Check with the wholesaler to see if he has any special purchases on various siding available. You may be able to pick up a good buy on a higher quality material at the same price of a low or middle grade of another product. If a substitute is adaptable don't be afraid to change. Keep an open mind. If you can get a better quality product on the exterior of your home you will add to the resale value in the years to come.

In checking with the people who are selling the products, if you show interest in what they are selling they will show interest in you and tell you about the shortcuts of installing their product. Above all, deal with a reputable company.

GUTTERS AND DOWNSPOUTS

If you want gutters and downspouts included on your carpenters' bids you must specifically ask for them. Some may refuse to put them in because sheetmetal subcontractors usually install them. Others may jump at the chance to install them. In any case it is an easy job and any one carpenter or sheetmetal man can handle it. There shouldn't be any need to hire an additional subcontractor just to install these units.

Are gutters really necessary? As one salesman told me, they are nice if you like ice buildup on your roof in the winter, or leaves clogging them in the fall. But they are not really as effective as you would like to believe they are. They tend to pour the rainwater in a few central locations, and if it rains too hard they overflow anyway.

If your property is properly backfilled and graded and you're not sitting on a high water table, chances are you may be able to get by without them. Also, if your house is designed with a three foot wide or larger overhang you don't need them. If the rain water can't get close to your foundation you have solved your problem.

To insure that no water will back up into your basement from the lack of gutters and downspouts check the water drip line on your grass. Buy enough drain tile to go half way around your house. Split the tiles in half and, with the curved side down, bury them in the ground along the water line. Keep the open side up and tilted away from the house. Fill the open end with crushed rock. The rain water hits the rocks and bounces away from the house. Decorate this crushed rock with shrubs. Select a variety of your choice, which should include a few that absorb plenty of water.

Gutters and downspouts are an important addition to a house that doesn't have an overhang or other basic protection. But if your basement is waterproofed and wrapped in polyethylene (as described earlier), if you have a low water table, and your property slopes away from the foundation, you have a good chance of being able to get away without installing gutters and downspouts until proven otherwise.

SOD

There are several ways to approach landscaping, and the final solution depends on the size of the yard and the length and width of the driveway. It is always best, though, to use a landscaping specialist.

There are many types of sod you can purchase, so be careful to get one best suited for your particular area and soil. Some sods require more water than others. About the best for any area is Kentucky bluegrass. It will stay green the longest with average watering.

Sod originally grown on peat may cause you fits in the future. It takes, on the average, more watering to keep green than cultured sod grown on regular soil. Laying sod is a backbreaking job and if at all possible, depending on the size of your lot, it may be advantageous to have the landscaping subcontractor install it. Also, if something does go wrong with the sod, and the subcontractor has installed it, the problems are his and not yours.

The average roll of sod is 1×6 ft and the price will vary depending on grade of sod, time of year, and whether the landscaper supplies the labor. Usually it will not exceed 30−35 cents a roll. Make sure you have at least four inches of rich topsoil before the sod is installed. It is important to the growth of the sod.

Another way to grass your property is considerably cheaper, but it takes a much longer time. This process is called *seeding*. It will require you to block off the yard so animals and kids stay away. Also, the ground will be exposed for a longer period of time depending on what season of the year you start seeding. It will probably be a great source of irritation to the average wife who has to put up with the kids and dirt. By this time you may wish you had chosen to use the rolls of sod even though you are watching fresh, beautiful grass shoot up from the seeds. It can be a trying period for the best of marriages.

SHRUBS AND TREES

Give yourself some time to look over the lot and home after you move in to determine exactly what you want in the way of shrubs and trees. Observe what your neighbors have done and get the names or pictures of shrubs and trees that you feel would enhance your exterior. Like everything else, these are personal choices and a great deal of time should be spent in making them.

Check with your sod man, if using one, to see if he can get you a good deal. Or see other landscapers and retail stores that sell shrubs for the products that you want. Shrubs usually go on sale around Memorial Day and at the end of the summer. Check to see what kind of coil you have, because some shrubs and trees grow better and live longer in certain types of soil. Landscapers who know their products will be more than happy to tell you about the right combinations of trees, shrubs, and soil.

BLACKTOP

Many landscapers are also in the blacktop business and, in order to get both jobs, they may offer you a package deal that can save you time and money.

Blacktop has all the advantages of concrete at about half the cost. The properly cared for blacktop will last just as long as concrete. If the blacktop does develop cracks, there are all kinds of blacktop patching kits that have entered the market in the past few years.

The average cost of installed blacktop is 38−43 cents a square foot compared to about 75 cents a square foot for installed concrete.

Today's blacktop can be mixed with dolomite, a white cement-like powder. When mixed with the blacktop, laid, and dried, it will firm up almost as hard as cement. An additional advantage of blacktop is that it doesn't add to the tax base of your property. In most areas, a concrete driveway is considered permanent and a tax is placed on it as it raises the value of your home. Blacktop usually doesn't. Check with your local tax assessor to make sure.

The area where the blacktop will be installed must be leveled and smoothed. Then two inches of crushed rock is installed and leveled for a base. Usually, three inches of asphalt is put over the crushed rock and packed down to a thickness of about two inches. You should allow five to seven days drying and hardening time before you drive or walk on newly installed blacktop. If the weather is very hot don't park anything heavy on the blacktop for up to ten days, or longer if needed. It is all right to drive over it to put your car in the garage but play it safe depending on the season. Test the blacktop with your foot. If it doesn't feel firm give it a few more days. Don't forget, you will be using your driveway every day for years to come. A few days wait in the beginning may add several extra years of use.

Concrete driveways are expensive, can add to your tax base, and may crack if your ground settles too much during the first year. Cement men told me that you should wait a good year or two, let the ground settle, and then put down a concrete driveway. But you can still have your cement men include it in their bid. Most people won't wait a year or two because they don't like the unfinished appearance of a crushed rock driveway.

The same principle of waiting applies when putting in the sidewalk. Many cement men will not guarantee their walks won't crack or sink because of the ground settling. Again, they prefer to come back the following year and do a lasting job.

In comparing the two, blacktop will give a little when the ground is settling, is less expensive to install than concrete, will not add to the tax base of your home, and is easy to maintain and repair if necessary. Like your home, whether you choose blacktop or concrete, take care of it and you will enjoy it for years to come.

8

Inside the Home

Admittedly, the outside of your dream house is important because it is visible to the eyes of all beholders. But the inside of your home takes on an even greater importance because this is where your friends will be. Besides, you'll be spending most of your time inside your home and you'll want it as pleasing and relaxing as possible.

So let us now look at the various ways of completing the interior of your home as we enter the finishing phase.

DRYWALL CONSTRUCTION

Most housing contractors today have gone into using a product called plasterboard or Sheetrock for wall construction. This is a drywall concept where individual sheets of dried plaster, either 4 × 8 ft or 4 × 12 ft, are nailed to the wall studs and ceiling joists to begin the finishing phase of the house. Drywall is more economical and any carpenter should be able to install it, or you may want to get bids from subcontractors who do nothing but install plasterboard. The plasterboard concept has been well accepted and can eliminate an extra subcontractor if you desire.

It is difficult to say what benefit drywall construction has over plaster walls during the time the home is being built. One can save time by using drywall construction because the drying process of plaster may take from 3 to 6 weeks to complete. Also, with the new plasterboard products used today, it is difficult to tell the difference between the two types of walls. Drywall is easier to install and simple to cut (Fig. 8-1).

The drywall concept, however, requires a seaming and taping application after the original plasterboard is nailed to the rough-in studs and joists. Some carpenters may do their

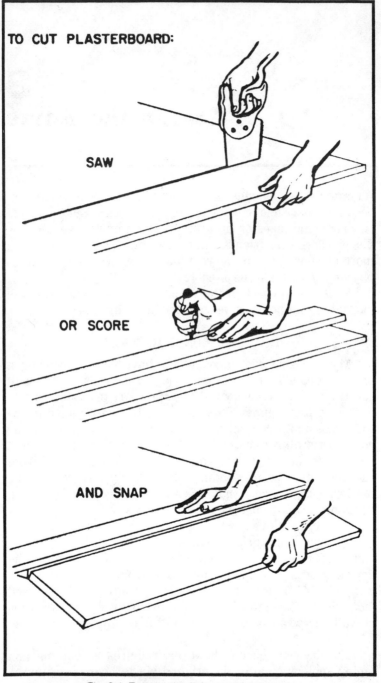

TO CUT PLASTERBOARD:

SAW

OR SCORE

AND SNAP

Fig. 8-1. Easy methods for cutting drywall.

own seaming and taping, or subcontract this portion of the job even though they included it in their bids. The job of seaming requires some skill in getting the proper thickness of cement. (See Fig. 8-2). The cement is first applied over all nails and seams where plasterboards join together. After the first very thin and even coat of cement is applied over the seams the seaming tape is placed directly over it. It must be laid on smooth and doublechecked to remove any air bubbles. Then both are permitted to dry.

Two more thin coats of cement are applied over the seaming tape. While seaming tape is used to join plasterboards and inside corners, a special perforated metal edge is used for outside corners. It is nailed in place and acts as a protective edge, otherwise the exposed plaster edge would eventually chip when accidentally hit. It is one way to childproof a home.

Once the outside metal edges are nailed in place several thick coats of cement are applied to them until the metal is covered. The covering cement is left with a semismooth surface. The whole process is a tedious job but nothing compared to the next step.

When all the coats are applied and dried they must be sanded down to a smooth surface with a fine-grain sand paper. Even if you are going to have your ceilings sprayed on, you should still have your applicator apply at least one coat of cement and seaming tape over the seams and one coat of cement over the nails. The ceiling should be seamed as evenly as the walls. This is necessary because it will give your sprayed ceiling a more even and consistent look and eliminate

Fig. 8-2. Drywall subcontractor finishes seaming to prepare walls for painting and paneling.

Fig. 8-3. An example of finished drywall construction.

any shadowy image when it dries. (See Fig. 8-3 for an example of finished drywall.)

If you are going to install paneling over any of the plasterboards, the applicator doesn't have to seam or tape that area. The labor hours involved in cutting and putting up the paneling in an area are about the same as seaming and cementing it. This means the labor cost should remain the same. Any extra cost, at this point, comes in buying the paneling. Part of this cost can be absorbed in the trimming of the painting bid or, better yet, doing the painting yourself.

If you are going to panel half way up a wall, the entire wall will have to be seamed and cemented and the carpenter will probably charge you an additional flat rate based on the number of panels he has to cut and install.

STUCCO AND PLASTER

Should you choose to have a stucco exterior or plaster walls on the interior, usually one subcontractor will give you a bid on both jobs. Here again you should get separate bids to leave your self open for alternatives. Above all, get references and take a look at some of their work. Three or four bids on each job should be enough to determine the cost of this portion of your home. If you feel you would like plaster walls instead of drywalls be sure to inform your carpenters before you get their bids.

Plaster walls differ from drywall construction in that the carpenters must nail a 48 × 32 in. piece of ⅜ in. Rocklath to the interior studs as they begin the finishing stage of the house.

Rocklath is similar to Sheetrock but has a rough surface. After the carpenters form the finished walls with Rocklath, the plasterers go to work. The mix a rough first coat of plaster and completely cover the Rocklath. The plaster fuses and drys against the rough surface of the Rocklath. The first coat should be permitted to dry for about 24 hours. The second coat is then applied and smoothed or left with a design depending on what you want the plasterers to do. Each coat is about ⅛ in. thick and the total wall will be about ¾ in. thick compared to a drywall of ½ in. thickness. The cost of Rocklath may be a little more expensive than plasterboard to cover the same area. The cost of labor to plaster the walls will be more expensive. There is probably a difference of $500 to $1000 to have plaster walls in your home. Let your pocketbook be your guide to determine which way you want to go.

A stucco exterior is beautiful, can be combined with any other exterior siding, and is usually good for the life of the home. It may also be very expensive. Still, you must see how this type of exterior fits into the overall appearance of your home. Don't rule it out just because of cost. In the long run it will pay for itself because the upkeep will be at a minimum. If you decide to stucco, the materials you purchase will be at a minimum cost as well. Notify your lumber dealers before you get their bids that you have decided your exterior is going to be stucco. If you are going to combine a particular siding with the stucco, inform the lumber dealer what that product is and how much of an area it is going to cover.

Since the only materials under a stucco exterior are the particleboard or plywood sheathing, asphalt paper (which goes over the sheathing), and a wire mesh screen (which is nailed over the asphalt paper), the overall price is kept low in comparison to other siding products.

The first coat of stucco is applied to the siding and held in place by the screen. Depending on the weather, it should dry for at least 24 hours. The second and finishing coats are then applied and smoothed or designed to your preference. At this point you may prefer to leave the stucco with its natural color, although many people will paint it with the color of their choice.

Compared to other siding, the price of stucco may balance itself out because although the material cost will be low the labor cost will bring it back in line. Since we didn't use either a stucco exterior or plaster on the interior walls, we don't have any way of determining how many hours it would take to complete an average job. From what I have been able to determine, by talking to other subcontractors and seeing how fast they can complete their portions of a home, I believe it doesn't take as long as one might think. You may want to use the comparative chart on how long it took to tape, cover the nails, and sand the plasterboard into shape to get a rough idea of an approximate time period.

SPRAYING THE CEILING

Depending on which subcontractor you get to seam and tape your plasterboard walls, that same individual should spray the ceiling if you choose to have this done. Even if your carpenter's bid includes this area, he may subcontract the job out. Or, if you get a special subcontractor to install the plasterboard walls, he should include the spraying in his bid. Whoever does it, it's still an easy job.

The quickest, easiest, and cheapest way to cover the raw plasterboard ceiling today is by blowing on a thick coat of a wet cement-like mixture. It is cheaper than painting because an average size house can have its ceiling sprayed within two hours. In addition, closets can be sprayed to eliminate the need for painting those areas. A small extra charge is necessary if you want it done. If you have the closets sprayed, stipulate that you want a flat coat of paint to cover it because the rough plaster coat will come off the walls when they are rubbed against either by you or the clothes in the closet.

If you request it, for a minimum charge the subcontractor will spray $1/8$ in. squares of aluminum into the wet cement. The chips will glisten and reflect light. It is very effective and adds to the total appearance of any room.

The average sprayed ceiling should not have to be covered or painted for at least 12–15 years. This would depend on how adequate a ventilation system you have in the house, how heavy a smoker you are, and the average amount of air pollution that exists in the area where you live. Contractors estimate that the average cost of spraying the ceiling is about ten cents a square foot.

There are certain rooms in your home where I recommend you don't have the ceiling sprayed. They are the work area in your kitchen, where the cooking and dishwashing is done, and the bathrooms, where humidity from showers may take its toll and eventually loosen the sprayed plaster from the ceiling.

A good two or three coats of semigloss paint with an oilbase in these ceiling areas is a must. But remember, always use a latex paint for the first coat to cover raw plasterboard. An oilbase paint will bubble up the paper covering of the plasterboard.

Spraying on the ceiling is a messy job, as you would expect. The cement is mixed on the spot, usually in a drum using a large electric mixer. Small amounts are poured into a handcarried, electric spraying machine and held about two feet away from the ceiling. The contractor swings the machine back and forth in a smooth motion making sure every area is covered. The cement must be sprayed in an even and consistent motion. It takes about 12 hours to dry. You should have some cheap plastic throws covering the floor so you don't have to waste much time in cleaning up afterwards. Also, any cement that gets on the walls can be scraped off easily before and after it drys.

STAINING AND PAINTING

When asking a professional stainer/painter for a bid, in addition to giving him your blueprints make sure he knows what is going to be stained, how much staining will be required, and what kind of wood trim and cabinets he will be working with.

Staining is a very time consuming and tedious job. A good stainer doing a good job must go through five separate steps to complete it. They include staining, sealing, puttying, sanding, and varnishing. Of course, the color of stain selected will be your choice but from there on the procedure is the same.

If you are trying to match paneling, carpeting, or anything else, then I suggest you get the help of a professional stainer. Also, for a clean, even look all through the home a professional is your safest bet. As you would soon see, a poor staining job will stick out "like a sore thumb." A good or great job should hardly be noticed after the newness of the home wears off. If you do use a stainer don't get any prestained woodwork, especially cabinets, from a supplier.

Make sure the contractor stains the kitchen cabinets, or any other cabinets, before they are put in their permanent place in the house. The plain, unfinished wood of the various cabinets will vary in color. With the cabinets standing side by side, the stainer can apply the stain, wipe it off to the desired shade, and get an even appearance across the various woods.

The hardest part of staining is wiping it to an even consistency. The best wood to work on is oak. It is clean and hard and holds its appearance. Raw pine, mahogany, and most other woods tend to blur or look a little muddy after the stain is applied.

Generally a stainer, if asked, will also give you a bid on the painting as a separate job. In any case, it is wise to get separate bids for you may decide to split the work or do part of it yourself. In this way you are not totally committed to one individual or company. Since this is a crucial part of the decor of your home, you should ask the stainer for references and see some of the jobs he has done. If you have no idea where to look for a stainer or painter, go back to the classified section of your newspaper or telephone directory.

How necessary is a professional painter? If you need to save money then I suggest you do it yourself. When buying paint buy the very best you can afford, for this is what you are going to have to live with and be proud of.

It is no easy job, but with good paint you can do it as well as any professional. Just remember, the first coat you put on raw plasterboard is always a latex waterbase, even in areas where an oilbase will be the second coat. Don't put the first coat on very heavily. A light coat will do because the surface of raw plasterboard has a tendency to bubble and flake if it gets too wet. An oilbase paint used as a first coat will cause nothing but bubbles, flakes, and problems.

If you can hold a roller or brush in your hand and enjoy selecting your favorite colors, you can be your own interior decorator and have a great feeling of self-satisfaction.

PANELING

The most popular sized paneling used in homes today are 4×8 ft sheets in thicknesses varying from $\frac{3}{16}$ to $\frac{5}{8}$ in. The most popular thickness is $\frac{1}{4}$ in. Remember one important fact when using paneling around an archway where a door will be hung. Be sure to order oversized doorjambs because the paneling can add up to a half-inch to the width of the wall by the doorway. The oversized doorjamb compensates for the added width of the paneling over the plasterboard.

If you plan to cut your paneling remember to predetermine how high up the wall you want to put it. If you want to go more than 36 in. and up to 48 in. high, you can get two pieces from each 4 × 8 ft panel with a minimum of waste. But if you only want to go up to a maximum of 36 in., you can get three sections out of a 4 × 8 ft panel. It's simple. Just cut your panel in three 32 in. sections and place it on the wall two inches from the floor. Then use a standard 2½ in. floor molding and 2½ in. cap molding over the panel. For example, a 12 × 12 ft room would only require four standard 4 × 8 ft panels instead of six standard panels.

The more area you panel the less you have to paint, but be sure to leave enough area open to paint or wallpaper so that a little redecorating will have an effect on your home corresponding to the mood you're in during any particular period. (See Fig. 8-4.)

Today's market provides the homeowner with a wide variety of paneling. While the original panels were made of wood or wood products, Masonite, a heavy paper-type paneling, has become popular and with good reason. Masonite has a wide variety of beautiful plush styles. But be aware that it does have one minor flaw. This type of paneling has a

Fig. 8-4. A fully paneled room with a partial view of wooden beams on the ceiling.

tendency to dry out and contract. This occurs no matter how perfectly the carpenters install it. The panels will separate by a hairline or more.

Depending on the pattern you purchase, you can get around this by using lattice (a strip of wood) placed over the seams where the panels meet. Properly stained lattice can enhance the beauty of the panels to the point of elegance if used in the right combination.

If you are interested in paneling check with your carpenter and lumberyard for their opinions and literature. Stick with popular local or national brands. These are the easiest to get and the manufacturers will back up their product. Should you or anyone else scratch them, there are many ways to repair paneling so that it is as good as new.

WOOD BEAMS

Along with paneling, another popular feature of home building today includes the use of beams on a flat ceiling in one or more rooms.

If the beams are on the blueprints they will be included in the bid from your carpenters. The installation of beams can be accomplished in a number of ways and good lumberyards have sample materials and literature on most of them. Most beams are actually false, relatively inexpensive, and can be put up in a short period of time.

The beams are placed on the ceiling after it is sprayed or finished. They are usually placed four feet apart at a point where the plasterboards meet. The beams help to cover up the seams, gives the ceiling a symmetrical look, and nails driven into the beams can go through the plasterboard directly into the ceiling joists for additional support.

The three basic beams used today are plastic or foam, solid wood, and rough sawn cedar for the rustic look. The plastic or foam beam has a prefinished appearance that looks false when it ages. It is sold by the running foot in 14 ft and 15 ft lengths and attached to the ceiling with a special glue. I have heard from reliable sources that the glue will stick nicely for about three years and then the beams will have to be reglued or bolted to the joists in the ceiling. This problem will also depend upon the amount of humidity in your home.

The solid-core beam is the most expensive to buy and can be purchased with a rough or smooth finish. Also, your stainer will include the beams in his bid if they are indicated on your blueprints. They are easy to stain because the type of wood used to form the beams will absorb the stain and usually discolor it to a shade a little darker than the rest of the woodwork. Don't be concerned as, more than likely, visitors will consider the discoloring natural and it will blend right into the design of the room. The solid beams will have to be nailed through the plasterboard and anchored into the supporting joists as this type is one of the heaviest you can use.

A rough sawn cedar beam is actually three cedar planks of lumber nailed together (after the wood is stained) to form a beam. The beam is nailed together from two sides with the enclosed side facing the floor. You also have a choice of having the beam even all the way around or the bottom side indented. Before the beam can be nailed to the ceiling, a strip of 1 × 4 in. pine, the inside width of the beam, is nailed in place. Both sides are nailed to the pine strip and to each end of the walls. The nails are recessed into the wood with a punch and woodputty is used to cover the nails. The overall effect should give it the appearance of the solid decorative beam at a less expensive price.

The thickness or size of the beam is up to your individual taste and pocketbook. The beam doesn't have to be very large. The average size, I would say, would be about four inches deep and four inches wide. This size doesn't overpower a room, but it stands out and adds distinctive boldness to the design.

FLOORING

Pick a floor, any floor, and in five to ten years you will probably be tired of it and want to change it. It's the easiest and the quickest way to redecorate your home today.

This doesn't mean you shouldn't ponder over your original choice. It does mean you should consider each room or combination of rooms individually so they can have their own distinctive personality. This will bring your home to life and give it additional warmth.

It is amazing what a piece of carpet can do for a newly constructed and decorated home with its bare floors still exposed. I can truthfully say I have always had a leaning in favor of carpeting and, since I became involved in this building project, it has been strengthened. In fact, we probably spent more time discussing flooring than any other single category of the home.

Today's carpet selections are fantastic. Not only are there expanded versions of traditional lines, but new innovations are reaching the stores every day. Shag carpets are still a popular style, carpets that look like area rugs are on the upswing, and even coordinated carpet and linoleum patterns are coming into their own. It could prove to be a popular mix and match opportunity for someone who is looking for an original approach to their dream home. Carpet is expensive but there are ways of cutting the expense down.

We bought the best juteback carpet we could afford to use in the areas of the home that will get the most wear and where we would entertain company. Also, we bought the best pad, 100 oz rubber with a waffle weave, for the plush feel underfoot. And we had it professionally installed. We paid for this by averaging it out against buying less expensive rubberback carpet for the bedrooms, one bathroom, and the kitchen. While juteback is difficult to seam and install by a layman, rubberback is very easy to install yourself, and once you get the hang of it you can do an average room in a few hours.

Who should install your carpet? There is no question in my mind that practically anyone can install their own rubberback carpet. The carpet manufacturers have done an excellent job in constructing their product to make it easy for the average person to install.

If you are considering carpet for your kitchen my advice is to get it. The kitchen carpets being produced today are as good as linoleum, as easy to keep clean, and very close in price. But remember, buy a good grade of kitchen carpet and save the difference by installing it yourself.

Juteback carpet is another matter. Where you can get by with double-face tape in seaming two pieces of rubberback carpet, you need more experience and an electric heatsealing handtool to melt the dried gluetape used in fusing two pieces of juteback carpet. Juteback carpet has to be stretched and doubled over a strip of smoothedge that holds the carpet firmly in place. Smoothedge is a thin piece of wood nailed to the edge of the floor with small preset nails sticking upward. When you are buying expensive and semiexpensive carpet, have the carpet company's mechanics install it. Here you should pay the installation charge. Consider it an insurance payment. If something should go wrong with the carpet because of poor installation the company will usually back it up.

Linoleum, slate, terrazzo, and ceramic tiles are all great products in their place. They have three specific advantages. They beautify the area they cover, protect the rest of the house from dirt and water, and accent or complement the flooring in the adjoining rooms. We chose to use linoleum in our entry way, mud room, and bathrooms. These areas, since they are close to outside doors, will take a beating from the dirt, mud, and water. The linoleum will act as a barrier for the rest of the home. Terrazzo, slate, and ceramic tiles are just higher priced versions of linoleum, but they add a more polished and plush appearance to the home.

Hardwood oak flooring is an expensive proposition and has to be put in during the construction of the home. Because of the multitude of flooring possibilities, hardwood flooring is almost impractical to use except where it is an integral part of the overall interior design of the house.

Before you start to build, you must have a good idea of what areas you intend to carpet and what areas you intend to tile. This must be known so that the lumberyard will also know what kind of finished flooring to include in their bid. Where carpet will be installed, a less porous material, such as particleboard, can be used. But where linoleum or similar products are used, and a paste has to be put on the finished floor, then plywood flooring should be used since the linoleum will adhere better to it.

The particleboard and plywood are cheaper and faster to install than hardwood floors. They are every bit as durable and dependable as hardwood floors.

Remember, it's your home and you must decide what combinations of flooring will reflect the image you are trying to project for your dream home. But don't get stuck with any permanent flooring that can't be picked up or recovered in the future. You may decide to change the personality of your home overnight.

CERAMIC TILE

Price is the largest factor in the category of ceramic tile. The most practical place for it is in the bathroom. It is probably the most well-known product for this purpose. Generally speaking, ceramic tile goes around the bottom four feet of the bathroom wall except for six feet around the bathtub-shower combination.

In the past, the chief competition was plastic tile used in less expensive homes. The difference between the two is the cost of the individual tiles. Also, ceramic tile installation is more expensive but in line with the value of the product. In my thinking, ceramic tile in the bathroom adds much to the total appearance of the home and its overall value.

The remainder of the exposed walls may be accented with matching or contrasting vinyl wallpaper or oilbase paint.

Both plastic tiles and ceramic tiles are put on the wall in the same manner. An adhesive is applied to the wall, then the tiles are placed over it in a series of rows. Plastic tiles with smooth edges are butted next to each other while the rough edges of ceramic tile don't allow for as close a fit. Thus, grout, a cement-like mixture, is placed between the butted ragged edges of the ceramic tile. After the grout is in place, and before it dries, wash away the excess with a sponge. The grout acts as a sealing agent and accents the tiles. After the grout completely dries, wipe the face of the tile with a dry cloth.

Ceramic tile is meeting new competition every day. Inexpensive one-piece drop-in shower and tub units, and Corfan wall sections are now being sold. The cost of these units may vary around the country, but if they become more popular and production increases the cost factor may change.

New wall coverings for bathrooms are coming out every day and it will behoove you to check with local retail stores and wholesale outlets for information on the latest products available. The entire building industry is constantly developing new ideas for our fast-changing society. The businesses involved in manufacturing building products are probably the most progressive in this country.

Who should you get to install ceramic tile? Actually it is very simple to install providing you have the correct ceramic cutting tools. If you have the time you may want to do the job yourself. Just buy the tile, paste, and grout. You can rent the tools. Mark off the area that you are going to cover with tile. Cover the entire area with paste, using a trowel. You should finish with parallel rows of paste.

As you apply the tile don't slide it, but twist it into place over the paste. The paste will spread out. Cap tile is used as a border over the regular square tiles. If the floor covering is going to be linoleum you should use a ceramic curved-base cove as the bottom row. The base cove must be installed after the linoleum is put on the floor. The linoleum and grout should fill in the open area between the linoleum flooring and the lip of the tile. If you are going to install carpeting on the floor, regular square tile can be used as the base tile.

The most important part of installing ceramic tile is getting the first row straight. Start at the floor level if carpeting is to cover the floor, or sparingly use the base cove as a guide to build a second row of square tiles. Be sure to work upward and outward instead of in a straight line. When all the tiles are in place mix the grout to the proper thickness and apply between the tiles. Let it set, but not dry, and wipe the excess from the tiles.

If this sounds like it may be more involved than it's worth, take your blueprints to a tile distributor to find the cost of the material and to locate a mechanic. The tile distributor will be able to inform you about mechanics looking for work. You can also check the classified section of your local newspaper or telephone book.

In our case, we went to a dealer who had the materials and installers and to individual installers. They all gave us bids for the entire job on both upstairs bathrooms. After comparing their bids, we found the dealer was by far the least expensive. And, best of all, the quality of work was what we hoped it would be. Check the man-hour work schedule to see how quickly a good installer can finish an average bathroom.

MIRRORS

Mirrors are expensive. It is important to consider what kind of mirrors you are going to have in your bathroom and throughout the house. If your home will have only one bathroom you should get a mirror attached to a recessed utility cabinet. If you have bathrooms back to back with parallel sink hookups you will only be able to insert one recessed wall cabinet between one set of wall studs in the adjoining wall. If you want another recessed cabinet in the opposite bathroom you may have to use a side wall or off-center the cabinet between a different set of wall studs in the adjoining bathroom. This should be determined before you finish your blueprints.

At this point you may want to consider using plate glass mirrors over your sinks or vanities. If you have vanities in the bathrooms you will have plenty of storage area for your supplies and you will be able to get by without a recessed wall cabinet.

Be sure the plate glass mirrors have predrilled holes in the four corners and that they come with special support screws. Check with your carpenter before he installs them to make sure he has the correct screws. When installing the mirrors make sure the screws are not completely tight. They must have a little play in them or you will risk cracking the mirror during installation.

If you are going to hang a full-length plate glass mirror on a hollow-core door, check with the millwork company or the lumberyard where it was purchased to find out where in the internal frame there may be cross supports that can be screwed into. This is necessary because you will have to tell the mirror company where the predrilled holes should be placed.

The same goes for hanging the mirror on the wall. If the mirror is large enough you may be able to drill directly into the wall studs for ideal support. You should not need special support screws if you can use the studs. Support screws will be necessary if the mirror is placed directly onto the plasterboard. Don't be confused or worried over this minor part. Good carpenters can do this job with hardly any problem at all.

There is a considerable mark-up on mirrors and here, as in every other case, you should make an attempt to get your contractor's discount. If you prefer recessed bathroom wall cabinets you may be able to get them included in your bid with your plumbing supplies, such as the sink, tub, bowl, vanity, pipes, etc.

APPLIANCES

The chances are that this venture will not be your first home and that you have most or all of the conventional appliances, such as a washer, dryer, range, dishwasher, freezer, etc. If you plan on taking any of these with you remember to get their exact measurements so they can be fitted into the blueprints.

If you are going to purchase new equipment get all of the written literature you can so that your plans can be adapted to the new pieces. Also, check to see if you have room to build or have freestanding room for new appliances such as a trashmasher, an inside gas or electric grill, a Vent-A-Hood, or a microwave oven.

Of course these appliances we are concerned with will fit into the kitchen, mud room, or basement. So the thing to do is first determine what kind of kitchen is wanted and how much room is needed. Remember men, this is usually the control tower of the home. Chances are the woman will be running the home from this one room. So give up early and save yourself a lot of heartaches. There is a lot of home left and, if you're lucky, your wife may save you one room just for yourself, just like mine did.

9
Coordinating the Construction

As you select your subcontractors remember to keep them down in number to an absolute minimum. The fewer you have to deal with the easier it will be to coordinate the building. Remember, you must coordinate the subcontractors, the inspections, and the material deliveries.

The materials should be selected before you start to build. This is a tough job because it involves picking out the raw materials, such as lumber, siding, shingles, etc., and the finished materials, such as carpets, lighting fixtures, paneling, color of the stain, etc. Chances are this is an area where your wife will dominate. Remind your wife about the budget and then let her make the selections. If the item will be displayed in the finished house she will want to select it. If it will be hidden she will allow you to make the final decision.

The important thing to remember is to have everything ordered in advance to allow the companies time to receive the stock. Give them six to eight weeks to fill your orders. They will be happy to hold your order for you and deliver it to the jobsite when you are ready.

As the time approaches to break ground for your home, don't let the butterflies or your wife get the best of you. If you have made it to this point, the rest is a continuation of gathering information to bring the materials and workmen together at their designated times. Keep in constant contact with all your subcontractors.

BREAKING GROUND

Let all your subcontractors and suppliers know when you are going to break ground. On the day you start meet the excavator on the sight and go over the area where the surveyor's stakes indicate the outline of the house on the property. He will then dig the basement to these points plus an additional two feet on the outside of each wall. This is needed so the outside brick walls can be waterproofed and backfilled. It is also easier for the cement men to work with the additional space.

The night before the excavator is to be there contact your cement man to make sure he has you scheduled. Remember, the weather can throw him off a day or two, so if you are unlucky enough to have bad weather it might delay you a little in the beginning.

If the cement men are ordering the blocks and the cement, which should be included in their bid, you have no worries. Otherwise, check with your brick and cement supplier and let him know what day the cement men are going to be starting their work.

As the cement men start to build the walls (see Fig. 9-1), notify your carpenters that they will be needed in about three to four days. Also contact your lumberyard and the company that sold you the steel beams and columns to deliver on a day both you and the carpenters agree on, weather permitting.

As the basement nears completion inform the carpenters that their time is near. Contact the electric company so they can run a temporary power pole next to the homesite to have power for the carpenters' electric tools. Also, contact the

Fig. 9-1. The block walls are partially built, signifying that it is almost time for the carpenters to begin their phase of construction.

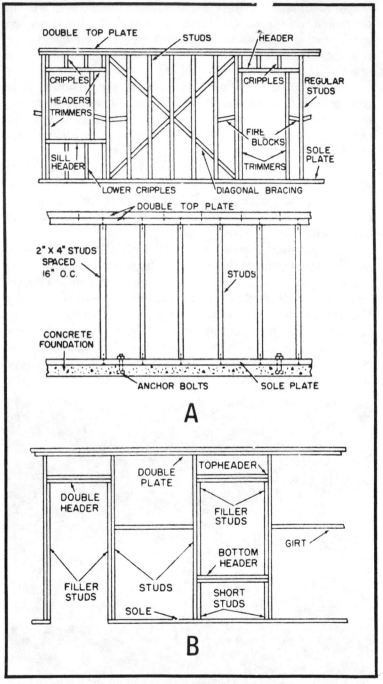

Fig. 9-2. Typical wall, window, and door woodframing.

excavator to tell him when the basement will be finished and when the carpenters will shore up the basement walls. The walls should set for about 24 hours before they are shored up. Usually, the carpenters won't cap the house until the excavator backfills the property.

During this waiting period, call in your waterproofer to cover the outside cement walls with hot tar. If you wish, you can then hang the polyethylene liner.

When your excavator returns and backfills against the foundation notify your carpenters. They should begin to set the beams and columns in place and start to cap the house. (See Fig. 9-2.)

ROUGHING IN

From this point use your carpenter as a guide to ordering your materials for delivery to the homesite. Check with him daily and tell him to give you a one day lead time for the additional material. A good lumberyard will send out enough material for two to three days work.

As the house continues to be framed up, there may be additional supplies that you may be required to buy or that the carpenters may pick up on their way to the home the next day. Our carpenters were extremely honest and showed us many ways of saving lumber and time—which means money. (See Fig. 9-3.)

At this point, after conferring with your carpenter, contact your millwork company or lumberyard and tell them when you will need the windows, frames, and outside doors. This is material needed to finish the rough-in stage. The carpenters may put these in before or after the other subcontractors have finished their part of the rough-in.

As the home gets closer to the completion of the rough-in stage contact your heating man, electrician, plumber, and telephone company. First check with your carpenter. He will tell you when the house will be ready for these subcontractors and their rough-in equipment. Give these subcontractors the exact date you will need them. As the rough-in is completed and the above subcontractors are working inside, the carpenters can be putting the shingles on the roof. (See Figs. 9-4 and 9-5.)

When you contact the other subcontractors, inform your roofing and siding dealer or lumberyard that you will be ready for the delivery of the roofing material on a specific day. They

Fig. 9-3. A view of the house during the frame-up period.

should be delivered by a truck with a hoist so the materials can be set directly on the roof, as this will be a time saver.

If you have a fireplace or brick exterior, be sure to contact your cement man and notify him that the rough-in of the home will be completed on a certain date. Remember to keep calling

Fig. 9-4. The house with rough-in completed and ready for doors, windows, and shingles to be installed.

Fig. 9-5. Windows are installed after rough-in.

or seeing the subcontractors if they don't show up and make them understand that they are an important factor in getting the home completed in a given period of time.

Again, if you are ordering any outside materials for the heating men, electricians, plumbers, or cement men, call the supplier and have the previously ordered material delivered on a date no later than the day the subcontractor is scheduled to start his work.

FINISHING STAGE

After the above subcontractors are done with their phase of the rough-in and the carpenters have finished the roof, the weather should be of little consequence in completing the house. If the carpenters are going to put the siding up on the outside as well as completing the inside they can shift around depending on weather conditions.

As the carpenters are finishing the roof order the siding, the insulation for the interior of the house, and the fire wall if you have an attached garage. As the carpenters continue to complete the house, check with them to determine when they want their next load of material. Again, this is a daily process and you must check with them to keep everything running smoothly.

If you are building with plasterboard for interior walls you will eliminate the plastering subcontractor. If you have hired one, then contact him as the house is being insulated and tell him when he will be needed. If a plasterer is called for, have your carpenters work to insulate and Rocklath the interior walls after they work on the exterior siding. The plasterers should work on the inside after the carpenters have sided the exterior.

Also, complete most or all of the exterior of the home before you plasterboard because the pounding on the exterior walls may pop the plasterboard nails after they are in place.

If you are going to stucco the exterior the carpenters should prepare it first. If the internal walls are going to be plastered, the same subcontractors would probably be doing the stuccoing. They should start on the outside when they finish the inside walls. If you are using an all-brick exterior follow the same procedure when contacting the subcontractor.

Always swing back and forth using your carpenters as a base. The finishing process of calling the subcontractors back to complete their portions of the home are identical to the rough-in stage, with a few additions.

Inform your cement man that the house is roughed-in and the basement is ready to take on the cement floor. If you have an attached garage, call the excavator to make sure the garage subfloor is leveled properly or to see if he will need to add more fill. If so, inform him that the cement men will be arriving on a certain day and that it is imperative for him to finish. The job will only take a few hours, at the most, so he could squeeze it into almost any workday.

As the concrete floor is poured in the basement and garage and the steps are being poured, the carpenters will stop working since anything they will do will raise dust that would eventually settle on the wet cement.

Upon completion of the cement work and the plasterboard or plaster walls, the heating men, electricians, and plumbers can come in to connect their equipment. Again, if you are bringing in any of your own material call the dealers and have them deliver it as soon as possible.

As the house develops in the finish stage, call the millwork company or lumberyard where your interior doors, trim, moldings, cabinets, and vanities are coming from to have them delivered. They will have to be stained before or after they are installed in their proper place. As in the other cases, call your stainer and let him know when he will be needed and how the house is coming along.

Again, coordinate the dates with the carpenters beforehand. In addition, when the vanities are stained, finished, and installed, call in your ceramic tile man, if you are using one, to install the tile. Provide him with the proper lead time.

If the finished flooring is almost completed and you are having linoleum installed, call the supplier and installer. Also call the telephone company to have them install their finishing equipment.

FINAL STAGE

When your heating and plumbing subcontractors are finished call the local gas company. They will send a crew to hook up the house to the local gasline, set the meter, and start the flow of gas into the home. Also, when the electrician has completed, call the electric company and they will transfer your service from the temporary pole to the main lines. Both the gas company and the electric company provide this service to you free of charge.

Depending on whether you are hooking up to a sewer system or installing a septic tank and cesspool, call your subcontractor and inform him that you are ready for his service. If it is a simple sewer hookup it will be a one day job. In any case, his work has little to do with the home itself and he can work independently except for a final hookup.

By the time you are this far into completing the home you will be a pro and know exactly what to do. Everything should fall into place as long as you keep the subcontractors working and the material arriving on time.

At this point the house needs to be decorated either by painting, carpeting, wallpapering, paneling, or a combination of any of these. Call the appropriate subcontractors giving the same lead time you gave the others. Don't take anything for granted and the home will be built accordingly.

As the house is being completed, the excavator should be brought back for the last time to grade the property and bring in any fill that is necessary to design the property to your specifications. When he is finished, your driveway installer, blacktopper, or concrete man should come in. Once these outside subcontractors are finished, the sod contractor should be contacted to complete the final exterior step in your building plans.

Building our house was no simple job, but it was a lot of fun. Personally, we found it one of the most rewarding periods in our life and I am sure, as you experience building your home, you will feel the same way.

Remember, this section doesn't include anything about calling the various inspectors because it was covered in another section. Be sure to call the inspectors at the designated times as it is part of the overall coordinating function.

CLEANUP

Even after all the work of planning, looking for subcontractors, high financing, and selecting the materials to go into your home, once the construction begins it will be up to you and your family to be your own cleanup committee.

Your subcontractors feel their responsibility is to build the home so their valuable time should not be spent cleaning it up. Usually, when the carpenters are working they will dump odds and ends out windows and doorways just to clear the way. They are the worst offenders because they have the biggest job to do. It should be up to you, or anyone you designate, to check out the house each night to see what you can do to make it easier for the workmen to start in again fresh the next morning.

During the week take an hour or two every day to remove lumber boxes, pop cans, books of matches, nails, etc., inadvertently left behind On the weekends, weather permitting, clean up around the outside as well as the inside. Determine what lumber you might be able to save and what has to be taken to a dump. Round up all the unused rough-in material, when rough-in is completed, and send it back to the lumberyard for possible credit if there is no further use for it.

It is amazing what your subcontractors will throw away that you might be able to keep for future use. Make two piles, one for unusable material and one for material you want to keep. As soon as your basement cement floor is poured and dried, take the reuseable material and store it in the basement. From personal experience, I figure we saved about $400 worth of material that I will be able to use in roughing-in and finishing the basement.

Don't consider this a dull or routine job and don't do it in a grudging manner. Look at it as a challenge that gives you the opportunity to inspect the house, see how the various subcontractors do their portion of the work, and gives you valuable insight into the total construction of the home.

Take it as a challenge and get the whole family involved. You will be surprised at how the members of your family will be willing to pitch in as it will make them feel like they helped in some way toward the completion of the home.

PROTECTING YOUR HOME

Of all the jobs I gave myself to do, the most distasteful one came after the workmen left for the night. Since you are your own contractor, it is your responsibility to keep the neighborhood kids away from the home to protect them from getting hurt, prevent the vandalism of your property, and see to the safety of the house.

It is a difficult and almost impossible job. If there is one phase that caused the most aggravation in building the home, this was it. There is no easy way to go about being the protector of your property. Don't be surprised at what you find in the attitudes of the neighborhood kids and even most of the parents. For some reason, everyone looks at new construction as fair game. In the long run the fair game is you.

Since kids are the main culprits, keep your eyes open for them. In particular, watch out for the girls more than the boys. For some strange reason the construction of a new home seems to bring out the strangest emotions in the girls. We found that boys explore the home, but generally do not cause any damage unless there are girls with them.

This is not a local phenomenon. I talked to a few contractors and they all seemed to agree that girls do create more of a problem, regardless of what area of a town a home is being constructed in.

There are a few solutions. You should normally be at the house at night to check out the work accomplished and to do some light cleaning. Your presence will keep most of the kids away. The boys and girls who usually don't cause any trouble will come around to visit with you. If you stay until it begins to get dark, a majority of your problems will be solved.

You can't be there around the clock, so it may not hurt to hire a protective agency to patrol the house three or four times each night. Call it term insurance. Once the people in the area know the house is being patrolled, they will have second thoughts about taking the chance of being caught on your property. The service cost us about $75 a month and was worth every penny. The patrols generally arrived between the hours of nine p.m. and five a.m., each day at a different time, to keep the kids and others off guard.

After the house is roughed-in, the windows in place, and the doors on, keep it locked. This is another point I can't stress enough. Always doublecheck the house before you leave to make sure it is secure for the night. If your house is unlocked and vandals do break in, some insurance policies may not cover the resulting damage.

Like everything else in this project it is up to you to protect the home while it is under construction. You must use a firm hand along with a gentle heart. Don't get upset by what you see going on. Let people know that if any of your material is missing, you will locate it and get it back. Always try to get friendly with as many kids as possible for they will eventually tell you who is taking what belongs to you.

INSURANCE

Here is where you will need the help of your professional insurance man. Tell him about your project. Inform him of your responsibilities and the cost of the home. You will need enough insurance to cover the total cost of the home starting from the day you break ground. It should be a builders risk type of policy that includes fire, windstorm damage, and theft. To play it safe in case anyone gets hurt on the job, carry liability insurance to protect yourself against a lawsuit. The chances of this happening are very slim but it could occur.

The insurance policy, all things considering, should not cost very much as you would only need it for a period of three to four months. When your construction is completed you can convert the policy to your normal homeowners insurance. It sounds easy and it was in my case. As long as you have a competent insurance man your problems in this area should be zero. Just doublecheck the policy when it is drawn up and make sure it is to your satisfaction.

10
Financing

Financing your house is a test of courage. If you want to build a house as much as we did then, somehow, you'll have to take the big step and see the officials at your local financial institution, be it a savings and loan or bank. Since we lacked the capital to build the home ourselves, I began to check around to see what was required by the various financial institutions for them to help us build. As you well know, without any money your plans will be useless.

The first step is to go into a financial institution to visit with a home loan officer and inquire as to what is needed to get the financial phase off the ground. A good place to start is the institution where you are currently doing business.

Chances are the loan officer will give you two forms and a booklet to complete. The forms are a *personal statement*, a statement of personal worth, and a *sworn construction statement*, which is used for the listing of bids from your selected subcontractors. The booklet is called a *home specification booklet* and will contain a written summary of the various areas that go into completing the home. Let's look at samples of these one at a time.

PERSONAL STATEMENT

As you can see, this statement (Figs. 10-1A and 10-1B) is designed to show the financial institution your true assets and liabilities. The institution requires this information to determine how much you can afford to repay them in monthly installments. It is also an indication to you of how big a house you might be able to afford.

At this point I would like to interject two points. First, try to build a home based on what you feel you will be earning two years from now. If you do, chances are you won't build a home

PERSONAL STATEMENT

To

Name (husband) _____

Name (wife) _____

Address _____

For the purpose of procuring credit from the above association, the following is submitted as being a true and accurate statement of the financial condition of the undersigned on the _____ day of _____ 19____

FILL ALL BLANKS, WRITING "NO" OR "NONE" WHERE NECESSARY TO COMPLETE INFORMATION

FINANCIAL CONDITION ON _____ 19 _____

ASSETS	Amount	LIABILITIES	Monthly Payment	Amount
Cash Accounts: Where deposited		Notes Payable to Banks (See Schedule)		
		Notes Payable to Others (See Schedule)		
		Other Accounts Payable (Itemize)		
Accounts, Notes Receivable (See Schedule)				
Earnest Money Deposit on Purchase				
U.S. Savings Bonds and Certificates				
Corporate Stocks and Bonds (See Schedule)				
Cash Value Life Ins. (See Schedule)				
Personal Property				
Business Equipment				
Automobiles				
Other Assets (Itemize)		Mortgage on Present Home (Mtg. holder)		
		Other Real Estate (See Schedule)		
Value of Present Home		**TOTAL LIABILITIES**		
Other Real Estate (See Schedule)		**NET WORTH**		
TOTAL ASSETS		**TOTAL LIABILITIES - NET WORTH**		

SOURCE OF INCOME		PERSONAL INFORMATION		
Base salary	$	Husband's Employer		
Overtime wages	$			
Bonus and commissions	$	Position held		No. of years
Dividends and interest income	$	Partner or officer in any other venture or other employment		
Real Estate income (Gross)	$	Wife's Employer		
Wife's income	$			
Other income - itemize	$	Position held		No. of years
		Single	Divorced	
		Married	Children (Ages)	
TOTAL	$	Age	Age of Wife	Other Dependents

Fig. 10-1A. Personal statement, front side.

that you would outgrow very fast. Second, talk to your boss. Tell him your plans and ask him for a letter stating what he feels you should be making two years from now. If you get this letter, attach it with the personal statement. You never know, it may be the item that convinces the institution you mean business.

SCHEDULE OF CORPORATE STOCKS AND BONDS OWNED

No. of Shares	Name of Company	Current Market Value

SCHEDULE OF OTHER REAL ESTATE

Property Location	Title in Whose Name	Value	Mortgage Amount	Mortgage Payment	Rentals Received	Mortgagee

SCHEDULE OF NOTES PAYABLE AND CONTRACT FOR DEED PAYMENTS

To Whom Payable	Security Pledged	Due	Monthly Payments	Balance

SCHEDULE OF LIFE INSURANCE

Amount	Name of Company	Beneficiary	Cash Surrender Value	Loans

SCHEDULE OF ACCOUNTS, NOTES RECEIVABLE

Maker or Payer	Security Pledged	Monthly Receivables	Date Due	Amount

GIVE NAMES OF BANKS OR FINANCE COMPANIES OR WHERE OTHER CREDIT HAS BEEN OBTAINED

Name and Address	Date	High Credit	Basis

The foregoing statements and details pertaining thereto, both printed and written, have been carefully read and the undersigned hereby solemnly declares and certifies that same is a full and correct exhibit of my/our financial condition.

Date _____ 19 _____   Signature _____

Fig. 10-1B. Personal statement, reverse side.

Now, discuss your personal statement with the loan officer. Get an approximate feeling of the institution's position on how much they would be willing to lend you toward the construction of your home. Remember, you are a novice and a financial institution is going to "pull in its horns" and not fully commit itself.

If you are fortunate enough to get some kind of a figure out of the loan officer, the next step is to get your plans developed and determine the breakdown by each subcontractor. Figure the cost of your home at about $20 a square foot. When discussing square feet, it should only include the above-ground living area, be it on one, two, or three levels. Therefore, if the institution feels you could handle a $30,000 mortgage your home should have 1,500 sq ft of above-ground living space, plus a basement. This is just a rule of thumb to go by.

You must consider what the property is going to cost, and if you own your present home what kind of a profit you can make when you sell it. It is difficult at this point to conceivably determine what you will be spending but take the following into consideration. We found most of our subcontractors stayed within the bids they quoted. The only subcontractors that varied were those that had to bring in additional material. On the whole, figure your home will cost about five to seven percent more than your original totals will estimate. This includes the extra lumber needed, minor changes, waste, and revising some of the original material.

Once you develop the plans the next step is to check out the sworn construction statement.

SWORN CONSTRUCTION STATEMENT

The sworn construction statement (Figs. 10-2A and 10-2B) is almost self-explanatory. It breaks down the home into the various categories needed to complete it. There is nothing mysterious about the form and its terminology. Remember, the banks and contractors have been using this form, or one similar, for years.

From here your fieldwork begins. You must locate several subcontractors and suppliers in each of the categories. It took me about three and a half months to make the final selections. Once your subcontractors are selected you must fill out the form by categories used and the total estimated cost. Break it down as finely as possible by dividing the material and labor costs as requested. Your subcontractors will be glad to give you the appropriate figures.

The columns to the right of the total cost should be completed as you pay out the subcontractors. The subcontractors should be paid in two installments. The first after the rough-in stage is completed; the second when the

Owner's name _____ Property at _____

	FURN. BY LAB., MAT. MEN & SUBCONTRS.	TOTAL COST	PD. BY OWNER	PARTIAL PAYMENT	BAL. UNPD.	LIEN WAIVER IN FILE	FULL PAYMENT
1. SURVEY							
2. EXCAVATING							
3. PILINGS – PIERS							
4. GRADING							
5. SOD & LANDSCAPE							
6. MASONRY							
7. CONCRETE							
8. WATERPROOFING – Drain Tile							
9. CHIMNEY & FIREPL.							
10. EXTERIOR WALL SURF. BRICK OR STONE							
OTHER							
11. LUMBER							
12. ROOFING							
13. INSULATION							
14. MILLWORK							
15. CABINET WORK							
16. HARDWARE							
17. SHEETROCK Plaster – Pntg.							
18. SEWAGE SYS. & TYPE							
19. WATER CONN. OR WELL							
20. PLUMBING							
21. HEATING							
22. AIR CONDITIONING							
23. SHEET METAL							
24. STEEL-IRON WORK							
25. FLOOR OR WALL TILE – MARBLE							
26. ELECTRIC WIRING							
27. ELECTRIC FIXTURES							
28. COUNTERTOPS							
29. LINOLEUM							
30. FINISHED FL. SURF.							
31. PAINTING & DECOR.							
32. CARPENTRY							
33. BUILT-IN APPLIANCES							
34. GARAGES							
35. PORCHES & DECKS							
36. SWIMMING POOL							
37. MISC.							
38. DETAIL ANY LABOR OF OWNER							
39. PERMITS							
40. LOT							
41. ARCHITECT							
42. DUE CONTRACTOR							
43.							
44.							
TOTAL							

(OVER)

Fig. 10-2A. Sworn construction statement, front side.

home is completed. In fact, your mortgage cannot be cleared until the last subcontractor is paid. Once the form is filled out the next step is to complete the home specification booklet.

HOME SPECIFICATION BOOKLET

I can only describe this 13-page booklet (Fig. 10-3) as a written outline of your blueprints. Since this booklet does get

A complete statement must be made giving the names and amounts due laborers, material men and sub-contractors furnishing work or materials on this project.

STATE OF
COUNTY OF _____ ss.

The undersigned, being first duly sworn, each for himself, as Contractor and Owner of the property

known as No. _____

City of _____ , State of _____ , deposes and says
that the foregoing are the names of all parties having contracts or subcontracts for specific portions of
the work on said property and building, or material entering into the construction thereof, and the a-
mounts due and to become due to each of said parties, that the items mentioned include all labor and
material required to complete said building according to plans and specifications; that there are no
other contracts outstanding; and that there is nothing due or to become due to any person for material,
labor or other work of any kind done upon said building other than as above stated.

The undersigned further deposes and says that no increase in the cost of construction will be made
under any circumstances without furnishing information on same to the
_____ with additional deposits to cover such increase; that in the event of any such
increase, no order or claims will be made on said company until such information and additional deposits shall have been completed; that the purpose of said statement is to induce said company to pay out

the proceeds of a loan of $ _____ secured by a mortgage on said property; and that, upon
payment of the specific unpaid items listed herein, waive all claims of priority to said mortgage and will
save said company harmless as to any other claims of priority of lien for any labor or material, furnished
or to be furnished, for completion of construction.

Subscribed and sworn to before me this _____ day of _____ , 19 _____

_____
                     Notary Public. _____ County.   (Contractor)

My commission expires _____
                                                                              (Owner)

IMPORTANT NOTICE: Any increase in cost from changes in construction or otherwise must be forth-
with reported to the   with additional deposits to cover
such increase.

Fig. 10-2B. Sworn construction statement, reverse side.

a little technical, the easiest thing to do is contact each of the subcontractors you have selected to build your home. Have them give you the information needed to complete the booklet. The samples shown in this book are directly out of the booklet we used in the completion of our home.

Actually you should get two of these booklets. One has to go to the loan officer and the other must be checked by the building inspector, along with your blueprints, before he will issue your building permit.

When the booklet is filled out take it along with your plot plan (Fig. 10-4), and blueprints, and sworn construction statement to the financial institution with which you are doing business. The loan officer will add the personal statement to it, have an inspector visit your building site to verify the property, and go into a mortage committee meeting to determine if he and the officers will accept, accept with reservations, or reject the mortage.

SPECIFICATIONS

The contractor shall provide all necessary labor and materials and perform all work of every nature whatsoever to be done in the erection of a residence for

_____

as owner, in accordance with these specifications and accompanying drawings. The location of the residence will be as follows:

_____

_____

GENERAL CONDITIONS

All blank spaces in these specifications that apply to this building are to be filled in. Items that do not apply are to be crossed out.

The general conditions herein set forth shall apply to any contract given under these specifications and shall be binding upon every sub-contractor as well as General contractor.

The plans, elevations, sections and detail drawings, together with these specifications, are to form the basis of the contract and are to be of equal force. Should anything be mentioned in these specifications and not shown in the drawings, or vice versa, the same shall be followed as if set forth in both, as it is the intent of these specifications and accompanying drawings to correspond and to embody every item and part necessary for the completion of the structure. The contractor shall comply with all health and building ordinances that are applicable.

EXCAVATION AND GRADING

The contractor shall do all necessary excavating and rough grading. The excavation shall be large enough to permit inspection of footings after the foundation has been completed. All excess dirt shall be hauled away by the contractor. Black surface loam to be piled where directed by owner for use in grading. Grade level shall be established by the owner who will also furnish a survey of the lot showing the location of the building. The finish grading, seeding, sodding and landscaping shall be done by the owner unless specified as follows: _____

_____

INSURANCE

The contractor will provide liability insurance and workmen's compensation insurance in full until completion of the building. Fire and windstorm insurance during construction to be provided by the owner.

EXTRAS OR CREDITS

Any deviation from these specifications or plans involving an extra charge or a credit must be agreed upon in writing between the contracting parties before the change is made. The contractor shall not take advantage of any discrepancies in the drawing and specifications. If any discrepancies are found they shall be referred to the owner or architect and be corrected before any contract is entered into. Dimensions shown in figures, on the drawing, shall take precedence over scaled dimensions.

— 1 —

Fig. 10-3. Home specification booklet (page 1 of 13).

CONCRETE FOOTINGS

Footings 2,000 PSI
Concrete footings for walls and piers shall be mixed in the proportion of one part cement,_____

parts sand and _____ parts gravel. Pit run gravel may be used if its proportions of sand and gravel are as called for. All aggregate shall be clean and sharp and free from organic matter. Coarse aggregate to pass 1 1/4-inch screen and to be retained upon a 1/4-inch screen.

Footings for walls shall be _____6_____ inches thick and _____4"_____ inches wider than wall on each

side: pier footings shall be not less than _____24_____ inches square and _____8_____ inches thick.

BASEMENT WALLS

Shall be of _____Concrete Block_____ construction, straight, plumb and level, and as shown on plans. All joints shall be struck flush on both sides. Beam fill as shown on plans. Basement walls

will or will not be waterproofed with _____2_____ coats of _____Tar_____

BASEMENT FLOOR

Shall be 4 inches thick. laid with sufficient slope to drain. If made in one pour, the mixture shall be the same as for the footings, troweled level and smooth. If made in two pours, the finish coat shall be one inch thick and shall be of a mixture of one part of cement to two parts of sand.

CEMENT WALKS AND STEPS

Cement walk from street curb line to front steps shall be _____ feet wide and _____ inches

thick. Walk from front steps to rear door step shall be _____ feet wide and _____ inches thick. The steps at the front and rear entrances shall be of wood, cement or brick construction as indicated on the plan. Concrete splash blocks shall or shall not be furnished.

CHIMNEYS

Chimneys shall be constructed of common brick with face brick top unless otherwise shown. The footings shall be concrete. The size of the flues shall be as shown. All mortar drippings shall be cleaned from flues leaving them uniform in size from top to bottom. All flues shall be properly lined with vitrified flue lining extending from the footing. Thimbles shall be placed as directed and cast-iron clean-out door shall be provided. If gas heating unit is used, the flue shall be circular vitrified

tile with bell top. or _____Class A Masonry Chimney_____
Fireplace. if any. shall be lined with fire brick and provided with a flue with an area not less than 1/10 of the area of the fireplace opening. Damper. ash dump and clean-out doors shall be provided. Circu-

lating type fireplaces. if any, shall be _____

Concrete hearth support to be fireproof. hearth floor to be _____Brick_____

Face of fireplace opening if masonry, shall be _____Brick_____

Mantel shelf if masonry, shall be _____Stone_____ Incinerator if any. shall be _____

BRICK WORK AND STONE WORK

All brick work shall be laid in cement and lime mortar, with all bricks well bedded and shoved into place. with both vertical and horizontal joints on straight lines. Joints to be of color selected by owner.

The price allowed for face brick, if any. is $ _____100.00_____ per 1,000. Any cost in excess of that amount shall be borne by the owner and any cost lower than this amount shall be credited to the owner. Lintels to be properly placed above all openings where masonry is shown above. Exterior stone work,

if any. shall be _____Red Brick_____

— 2 —

Fig. 10-3 (cont). Home specification booklet (page 2 of 13).

TILE WORK

The contractor shall furnish and set all tile, if any, in a neat and workmanlike manner. The tile shall be as follows: _____ Ceramic _____

Towel bars, grab bar, paper holder, soap dish or other fixtures, if furnished by tile contractor, shall be _____ Ceramic _____

CARPENTER WORK

The contractor shall and will provide all necessary labor and perform all carpenter work of every nature whatsoever to be done. He shall lay out all work and be responsible for all measurements, and keep a competent foreman in charge. All work shall be done in substantial conformity with the plans and specifications or any variations, changes or amendments thereof that have been approved in writing by the contracting parties

GIRDERS

Girders or supporting beams shall be as required by the size of building, and as shown on plans, all to be __6"x10" Const. Fir.__ Columns to be __6"x6" Fir.__ of size shown on plans.

JOISTS

First floor joists to be   2" x __10__ ", placed __16__ " on centers, Grade __N.C. Const. Fir.__

Second floor joists to be 2" x __10__ ", placed __16__ " on centers, Grade __N.C. Const. Fir.__

Ceiling joists to be   2" x __6__ ", placed __16__ " on centers, Grade __N.C. Const. Fir.__

Rafters to be   2" x __6__ ", placed __16__ " on centers, Grade __N.C. Const. Fir.__

Collar ties to be   2" x __4__ ", placed __32__ " on centers, Grade __N.C. Const. Fir.__

Valley and hip rafters to be 2" x __8__ ",   Grade __N.C. Const. Fir.__

STUDDINGS

Studdings shall be sized 2x4's, spaced 16" on centers, single plate on bottom and double plate on top of each wall or partition, 2x4's shall be doubled around all openings. Outside studdings shall be __Const. Fir.__ Inside studdings shall be __Const. Fir.__

BRIDGING

First and second floor joists shall have one row of 1x3 beveled bridging for all 14' spans or less. All spans over 14' shall have two rows, all nailed securely to joists at each end.

SUB-FLOORING

Sub-flooring shall be laid diagonally with __5/8" Plywood__ , securely nailed. All joints shall be made on joists.

— 3 —

Fig. 10-3 (cont). Home specification booklet (page 3 of 13).

SHEATHING

Outside wall sheathing shall be __1/2" Builtrite Nailbond type only or 1/2" Plywood__ Roof sheathing shall be __5/8" Plywood__

WALL SIDING

Siding, if any, to be __Aluminum__

ROOF SHINGLES

Shingles for roof to be __240# 3-Tab Sealdown__

laid __5__ inches to weather using galvanized nails. _____

SIDEWALL SHINGLES

Sidewall shingles, if any, to be __——__

INSULATION AND PAPER

Sidewall insulation to be __3"__ Top floor ceiling insulation to be __6"__

Building paper under shingles to be __15#. Starter Roll to be 50#__

Building paper over sheathing to be __15# Felt Tip__

Building paper between sub-floor and finish floor to be __Red Rosin 30#__

OUTSIDE FINISH

All lumber required for outside finish shall be __Redwood__

WINDOW AND DOOR FRAMES

All window and outside door frames as shown on plans shall be of sound clear pine, free from objectionable defects. Outside casing __1 1/4"__ thick. Door sills shall be __1 1/2"__

Assembled basement sash units, if any, shall be __Standard Size__

Assembled window units, if any, shall be __S —— Millworks__

Assembled door units, if any, shall be __S —— Millworks__

WEATHER STRIPPING

Shall be __Plastic__

WINDOWS, STORM SASH AND SCREENS

All windows and sash shall be __1 3/8"__ thick of pine, as shown on plans.

Screen cloth shall be __——__ mesh __——__ wire. Glass to be __——__

Combination storm sash and screen units, if any, shall be __S —— Millworks__

— 4 —

Fig. 10-3 (cont). Home specification booklet (page 4 of 13).

FINISH FLOORS

Finished floors in Living Room and Dining Room to be _5/8" Partical Board - Carpet_

Finished floor in Vestibule to be _Exterior Plywood - Inlay Tile_

Finished floors in Bedrooms to be _5/8" Partical Board - Carpet_

Finished floor in _____ to be _____

Finished floor in Kitchen to be _Inlay Tile over Exterior Plywood_

Finished floor in Rear Entry to be _Inlay Tile_ Porch fl. to be _____

Finished floor in Bathroom to be _Inlay Tile_ Fin. fl. in Toilet to be _Inlay Tile_

Linoleum sub-floor lining to be _Exterior Plywood_

Flooring in Attic of house to be _____ Flooring in Attic of Garage to be _____

All hardwood floors shall be properly nailed and machine-sanded to a smooth, even surface. Floors under linoleum shall be securely nailed with coated nails and machine sanded.

INSIDE FINISH

Trim in the Living Room, Dining Room and Vestibule shall be _Oak_

Trim in _____ shall be _____

Trim in the Kitchen and Rear Entry shall be _Oak_

Trim in the Bedrooms, Bathroom and Hall shall be _Oak_

All trim shall be freshly cut and sanded at the mill.

Base to be _Oak 2 1/2"_ Casings to be _1 member_

DOORS

All of the inside doors shall be 1-3/8" thick as follows: _Oak - Hollow Core_

The front door shall be 1-3/4" thick of _Oak - Solid Core_

The remaining outside doors shall be 1-3/4" thick of _Metal Door Solid Core Back Outside Garage_

Provide scuttle door to attic and plumbing access door.

Provide combination storm and screen doors for all outside doors. (Front Door)

— 5 —

Fig. 10-3 (cont). Home specification booklet (page 5 of 13).

JAMBS AND CASINGS

All inside jambs shall be 3/4" thick, and of kinds specified above. Casings shall be 2 1/4" " thick

of _____Oak_____ design. __Narrow Line_____

STAIRWORK

Stairs leading from first to second floor shall be as shown on plan with ___14___ risers and

___12___ treads and ___1___ wall rail. Basement stairs shall have ___12_____

risers and ___---___ treads and ___1___ wall rail. _____

Three stair horses shall be provided for each stair and shall be ___2___ " x ___12___ ".

Wall stringers shall or shall not be housed.

CABINET WORK

Built-in Medicine Cabinet in bathroom, size _____none_____

Kitchen cabinets shall be of Oak Veneer placed as shown on plan. Same shall or shall not have
3/8" Lip
Plywood backing and doors shall be 3/4 inches thick, shall be panel doors, shall be flush doors,
and shall or shall not be rabetted. Cabinet drawers and doors shall or shall not have lips. If catalog
3/8"
design is selected by owner, same shall be _____

_____

Kitchen counter tops to be Formica Splash back to be Formica

Mantel shelf (if shelf only), shall be _____ Design No. _____ Kind of wood _____

Mantel shelf and facing for fireplace opening, if wood, shall be Design No. ____ Kind of wood _____

_____

Other cabinet work, if any, shall be as follows: Bathroom Vanities _____

_____

Clothes Chute, if any, shall be ___metal_____

CLOSETS

All closets shall have the necessary hook strips with hooks and one shelf and one clothes rod.

STORAGE ROOMS

Provide one _____ x _____ storage room in basement, using _____

with _____ shelves 18 inches wide. _____

— 6 —

Fig. 10-3 (cont). Home specification booklet (page 6 of 13).

LATHING PLASTERING AND GROUNDS

Plaster **base** shall be gypsum lath, structural insulating lath, or metal lath and shall be applied in

accordance with the manufacturer's directions. Wood lath, if used, shall be _____

All plastering to be two coat work with _____

_____

All corners and angles must be finished plumb, straight and true, and all surfaces smooth and flush with grounds. The plasterer is to point up and finish all defects after carpenters are through if necessary and remove from premises all rubbish pertaining to his work and leave the building broom clean. Keene's cement will or will not be used in bathroom up to height of chair rail. All exposed plaster corners to be provided with metal corner beads, inside corners to be reinforced. Ceiling cove with one break at ceiling or wall shall or shall not be furnished. Plaster grounds around all inside

door openings are to be _____ . Temporary heat, if necessary, shall be furnished by_____ .

Plastering in the basement shall consist of the following: _____

_____ Plastering in garage shall consist of the following: _____

DRY WALL

Dry-Wall, if any, shall consist of ____ 1/2" Gypsum Board _____

All exposed corners shall or shall not be reinforced with ____ Bead _____

Ceiling cove, if any, shall be _____

Joints shall be finished by _____ Tape and Fill _____

INTERIOR PAINTING

All woodwork to be carefully cleaned of finger marks, stains and other defects before any oil, filling, paint or varnish is applied, and all rough spots to be sand-papered smooth before being filled or finished, and all nails and brad holes to be filled with colored putty to match color desired. Finish to consist of the following:

Living Room _____ X _____ Dining Room _____ X _____

Kitchen _____ X _____ Rear Entry _____

Bedrooms and Halls ___ X _____ Vestibule _____ X _____

Bathroom _____ X _____ Basement _____

Radiators _____ Floors _____

DECORATING

Interior wall decorating shall be as follows: _____

_____ _____

_____

EXTERIOR PAINTING

All exterior woodwork shall have _____ coats of pure lead and oil paint of colors to be selected by owner. All sash and trim to be neatly traced. All knots and other defective work to be shellacked, and all nail holes to be puttied before applying last coat. Roof to have a brush coat of creosote stain

if unstained wood shingles are used. All exposed sheet metal shall have _____

_____

— 7 —

Fig. 10-3 (cont). Home specification booklet (page 7 of 13).

STUCCO

All stucco, if any, shall be three coat work over _____ gauge _____ metal lath. The dash coat shall be selected as to color and kind by the owner. Put waterproof paper on sheathing under the metal lath. All work to be done in a neat and workmanlike manner.

HARDWARE

The contractor shall furnish all rough hardware and garage door hardware if garage is included. The amount to be allowed for finish hardware is $ _100.00_____. Any cost in excess of that amount shall be paid by owner and any cost lower than this amount shall be credited owner.

SHEET METAL

Contractor shall and will provide all necessary labor and materials and perform all sheet metal work of every nature whatsoever to be done, including gutters under all eaves with suitable conductors. All joints to be well soldered and securely fastened, and all work to be done in a neat and workman-like manner. Gutters to be _____ gauge _____ metal _____ type. Down spouts to be _____. Proper _____ flashing shall be provided wherever necessary. Valley flashing, if any, shall be _____ metal _____ inches wide. Clothes chute, if shown on plan, shall be lined with _____

HEATING

Contractor shall and will provide all necessary labor and material and perform all heating work of every nature whatsoever to be done, including all plumbing and electrical connections, in the installation of a heating plant of sufficient size to properly heat all parts of the house in the coldest weather, and to put same in full operating condition.

WARM AIR SYSTEM: If gravity type or blower type of hot air heating is to be used, same shall be installed according to the code of the National Society of Heating and Ventilating Engineers, and in all cases shall conform to local ordinance. The heating plant shall consist of the following:

Heil Gas Forced Air Furnace 135,000 BTU Input 00GL 135 AE - 1c
_____ Gas Forced Air Space Heater 75,000 BTU Input
above ground.   Model GU 75 MD - C for Basement

AIR-CONDITIONING SYSTEM: If any, shall be _Heil Gas Air Conditioner_
_37,000 BTU Input Model No CU37CQG1_____

_____

OTHER HEATING SYSTEM: If hot water, steam, or any other heating system is to be used, such installation shall consist of the following: _____

_____

GAS or OIL: If an oil or gas burner is to be used, such installation shall include the following:

_____

_____

— 8 —

Fig. 10-3 (cont). Home specification booklet (page 8 of 13).

98

PLUMBING

Contractor shall provide all labor and material and perform all plumbing work of every nature whatsoever to be done including items under "Built-Ins" if specified.

All plumbing shall be properly installed and all connections thoroughly tested and shall be installed according to local ordinances. Hot and cold water connections shall be made with bath tub, shower, lavatory, kitchen sink and laundry tray. Water connections shall be made with water main in the street, sewer connection shall be made with sewer in the street, gas connections shall be made with gas main in the street, all to be paid for by the plumbing contractor. Equipment under this plumbing contract requiring electrical connections shall be furnished and put into working operation by the plumbing contractor. All meters shall be paid for by owner. All piping to be according to local ordinance.

Plumbing fixtures shall be as follows: _____

____ Bath Tub _____ Color _____

____ Shower over Tub, if any shall or shall not have curtain rod but no curtain _____

_____

_____

____ Water Closet _____ Color _____

_____

_____

____ Lavatory _____ Color _____

_____

____ Kitchen Sink _____ Color _____

_____

_____

_____ Water heater with _____ gal. storage tank. Laundry tray _____

____ Rain leaders shall or shall not be connected to sewer. _____

_____ Lawn faucets, _____ Gas openings, _____ Floor drain, _____ Gas vent,
Towel bars, grab bar, paper holder, soap dish or other fixtures, if furnished by the plumbing contractor, shall be _____

BUILT-INS

Contractor shall provide all labor and materials for the installation of the built-in equipment herein specified.

Subcontractor shall connect with water, gas or electric supply the equipment he installs and shall put it into working operation. The incinerator, if any, and the fans, if any, shall be vented.

Garbage Disposer _____	Incinerator _____
Built-in Oven _____	Counter-top Range _____
Hood and Fan _____	Bathroom Fans _____
Shower Door _____	Tub Enclosure _____
Bath Heaters _____	Room Air Cond. _____
Refrigerator _____	Freezer _____
Dishwasher _____	Clothes Washer _____
Clothes Dryer _____	Comb. Washer-Dryer _____
Television _____	Radio-controlled Gar. Dr. Operator _____

Fig. 10-3 (cont). Home specification booklet (page 9 of 13).

ELECTRICAL WORK

Contractor shall provide all necessary labor and material and perform all electrical work of every nature whatsoever to be done. All work to comply with local ordinances. Provide wiring outlets as per following schedule. All outlets to be placed as directed by owner and shall consist of the following:

WIRING SCHEDULE

SYMBOLS	OUTLETS AND SWITCHES	LIVING	DINING	KITCHEN	FML. ROOM	HALL	STAIRWAYS	BEDROOM	BEDROOM	BEDROOM	CLOSETS	BATHROOM	LAVATORY	SHOWER	BASEMENT	LAUNDRY	UTILITIES	ENTRANCES	REC. ROOM	GARAGE	YARD	PORCHES	ATTIC	OFFICE	TOTAL
○	Ceiling Outlet	1	7	3	2		1	1			3	2	1		10							2		1	42
-○	Bracket Outlet																								
⊖	Convenience Outlet	8	5	9	5			3	3	4		2	1		1										45
⊙	Floor Outlet																								
▲	Special Purpose Outlet																								
⊖.	Range Outlet		1																						1
⊗	Ventilating Fan Outlet		1									3													4
⊖ₐₚ	Convenience Outlet (weatherproof)																			2					2
⌐E	Radio Outlet																								
-⊖	Clock Outlet		1																						1
○	Illuminated House Number																								
S	Lighting Switch	2	2	1		1		1	1			4	1			2	2	1						1	19
S₃	Three Way Switch				1				1						1										3
Sₚ	Switch and Pilot																								
Sₓₚ	Switch (weatherproof)																								
⊘	Bell and Transformer																		1						1
-|-	Circuit Breaker												1												1
◁	Telephone outlet	1		1	1		1	1	1						1										7

RECAP OF OUTLETS

_42_ Ceiling outlets _____ Bracket outlets _45_ Double wall plugs _____ Floor outlets
_19_ Switch outlets _3_ 3 way switch outlets _____ Switch and Pilots _2_ W proof switches
_____ Clock outlets _____ Radio outlets _1_ Bell and Transformer _7_ Telephone outlets
_4_ Fan outlets _1_ Range outlets _____ Yard outlets _____ Heating outlets

The electric entrance service shall be three-wires with __150__ ampere capacity, installed in accordance with the National Electrical Code or the local ordinance, whichever applies. The type of switches shall be __Quiet Type Contact__.
The electrical contractor shall include items specified under "Built-Ins".

ELECTRIC FIXTURES

Electric light fixtures to the value of $ __400.00__, including installation of same, shall be furnished by contractor. Any cost in excess of this amount shall be paid by the owner and any cost lower than this amount shall be credited owner. _____

— 10 —

Fig. 10-3 (cont). Home specification booklet (page 10 of 13).

WINDOW SHADES

The amount of $ 50.00 is allowed for window shades. Any cost in excess of this amount shall be paid by owner and any cost lower than this amount shall be credited to the owner.

CLEANING UP

The contractor shall ~~or shall not~~ remove all debris from the premises when the job is completed. The contractor shall ~~or shall not clean~~ all window glass when job is completed.

_____

_____

GARAGE

Foundation __2,000 PSI__ Floor 1-2-4 Mix __3,500 PSI__ Apron 1-2-4 Mix __3,500 PSI__ Thickness __4"__

Sheathing __5/8"__ Wall Covering __Aluminum__ Roof Shingles __Asphalt 240 lbs. 5 Tab__

Service Door __2'8" Pine__ Driveway __Blacktop__ Electric Wiring __Yes__ Painting_____

Garage Door ~~shall or~~ shall not be the hinged type _____

Garage Door shall ~~or shall not be~~ the overhead type _____4 - Section_____

Garage Door shall or ~~shall not~~ be mechanically operated _____

_____

_____

FIREPLACES

Fireplace, if any, shall be as shown herein. Specify type of mantel on page 2 or 6.

MISCELLANEOUS
(Describe items if any, not mentioned in specifications or shown on plans.)

_____

_____

_____

_____

_____

_____

_____

_____

_____

_____

_____

_____

_____

_____

_____

— 11 —

Fig. 10-3 (cont). Home specification booklet (page 11 of 13).

BUILDING CONTRACT

WE THE UNDERSIGNED, have read the foregoing specifications, prepared in duplicate, and accept them as correct and hereby acknowledge receipt of one copy for each party hereto.

THIS AGREEMENT, made this ____1 st____ day of ____May____, 19_76_, by

and between ____John Doe_____, hereinafter called

the Contractor, and ____Joseph Smith_____, hereinafter called the Owners, WITNESSETH:

That the Contractor and Owner for the consideration hereinafter named agree as follows:

ARTICLE I. The Contractor agrees to provide the materials as specified and to perform all the work shown on the drawing and subscribed in the specifications and to do everything required by the General Conditions of the Contract, the Specifications and the Drawings.

ARTICLE II. The Contractor agrees that the work under this contract shall be substantially completed the ___31 st___ day of ___October, 1976_____ .

ARTICLE III. The Owner agrees to pay the Contractor in current funds for the performance of

the contract ____XXXXXXXXXXXXXXXXXXXXXXXXXXXXXXXXXXXXX____ Dollars ($ __XXXX__), subject to any additions or deductions to the General Conditions of the Contract agreed upon in writing, and to make payments on account hereof upon presentation of proper lien waivers, as the work progresses and as follows: ____XXXXXXXXXXXXXXXXXXXXXXXXXXXXXXXXXXXXXXXXXXXXXXXXXX____

_____

_____

_____

_____

It is agreed that the completion of the work covered in this contract is contingent upon strikes, lockouts, delay of common carriers, laws or government regulations or any other circumstances or conditions beyond the control of the contractor.

ARTICLE IV. The Contractor and the Owner agree that the General Conditions, the Specifications and the Drawings, together with this agreement form the contract, as if hereto attached.

The Contractor and Owners for themselves, their successors, executors, administrators and assigns, hereby agree to the full performance of the covenants herein contained.

IN WITNESS WHEREOF, They have executed this agreement the day and year first above written.

____Joseph Smith____	Contractor ____JOhn Doe____
(Witness)	
____Joseph Smith____	Owner ____John Doe____
(Witness)	

Fig. 10-3 (cont). Home specification booklet (page 12 of 13).

Authority for Change in Contract

By mutual agreement between . . .

_____ (Contractor)

and _____ (Owner)

the following changes, alterations, additions or substitutions are to be made in the work or materials called for in the plans and specifications which are a part of the Building Contract signed by the above named parties.

on _____ 19 _____

_____

_____

_____

_____

_____

These changes will (cost or save) $ _____ which amount will be added to or subtracted from the next regular payment, or if agreed upon, at the time of final payment, under the terms of the above mentioned contract.

Contractor _____

Owner _____

_____19 _____

*Notify Mortgagee of any additions to,
or deductions from the contract price.*

Fig. 10-3 (cont). Home specification booklet (page 13 of 13).

If the loan is rejected find out why. Usually it's because the institution doesn't feel you will be able to build the home for the amount quoted, or if you do you won't be able to make the payments. At this stage of the game, it will be up to you to convince the loan officers that you can do it and that you will be earning enough when the time comes to make the payments.

If the loan committee decides you can build the home and that you can afford it they will tell you to go ahead. The loan officer must confer with you to determine how much of an interim loan you will need to pay the subcontractors during

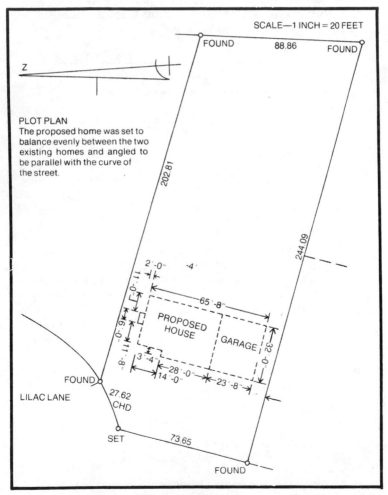

Fig. 10-4. Plot plan to be included with the home specification booklet.

various stages of the building program. Generally, the institution will be willing to go for half the total cost of the home. The institution will set up a loan, you will be given the money as you request it, and you will be charged interest only for the amount used and the period of time it is on loan. This is the easy way.

Why do you need this interim financing? Because most subcontractors work on a two-pay basis. Since all subcontractors at one time or another work for contractors, besides themselves, they understand how financial institutions handle the payouts. They want one payout at the completion of the rough-in stage and the second when the home is finished. Their payments generally coincide with the percentage of work they do that is necessary to complete each of the stages. Usually they will ask for half payment at rough-in and half at the finish.

I happen to have fallen into a third category which took some time to work out. The institution I used granted the loan but wouldn't give me the interim financing. They told me they would give me a note that I could take to a second institution, guaranteeing the second institution that they will lend me the full amount requested, repay the interim financing, and make final payments to subcontractors that have money coming to them. This note is called a *commitment letter*.

COMMITMENT LETTER

The financial institution will usually charge you a small percentage of the requested mortgage for the commitment letter, but the charge is refunded after the home is built, the subcontractors are paid in full, and the final mortgage papers are signed. If you get a commitment letter and you are not able to complete the home the charge is forfeited.

If you fall into this category, take the commitment letter in your hand and go to another financial institution where you are known or where someone may be able to introduce you to a financial officer. All you really need is a simple loan and you should approach it in this respect. When you meet the new loan officer show him the commitment letter. Inform him of your present financial situation, what kind of collateral you have to put up for the note, and the status of your present home. He will want to know what the present mortgage is, how much you feel you can sell it for, and what you may be able to clear out of the sale.

RECEIPT AND WAIVER OF MECHANICS' LIEN RIGHTS

N. B. It is important that the following directions be closely followed as otherwise the receipt WILL NOT BE ACCEPTED.

1. This is a LEGAL INSTRUMENT and must be executed accordingly by officers of corporations and by partners of co-partnerships.

2. It is important that ALL the blanks be completed and that the AMOUNT PAID BE SHOWN.

3. If payment is not in full to date, so state, SHOW UNPAID BALANCE, and strike out last three lines.

4. A receipt similar to this or legal waiver of lien rights will be required for all plumbing, heating and plastering material, etc.

5. NO ERASURES OR ALTERATIONS MUST BE MADE.

_____ , 19____
(DATE)

The *undersigned acknowledges having received payment of*_____

_____ _ DOLLARS ($_____)
(Amount Paid)

*from*_____ *in full payment of*
(Name of Payor)

*all*_____
(Kind of Material or Labor)

*by the undersigned delivered or furnished to (or performed at)*_____

_____

_____
(Street Address or Legal Description)

and for value received hereby waives all rights which may have been acquired by the undersigned to file mechanics' liens against said premises for labor, skill or material furnished to said premises prior to the date hereof.

PLUMBING, HEATING and PLASTERING   *By*_____
contractors must sign statement on reverse hereof   _____
(TITLE)

*Address*_____

Fig. 10-5. The lien waiver.

Again in our case, in order to secure the loan we had to use the commitment letter, sign an unrecorded second mortgage on our old home, and pledge a small amount of savings to act as collateral for the note. It sounds like a lot but all it amounted to was a lot of paperwork to protect both parties involved. If I had to do it again I would.

Our loan was approved with the security backing it up. The amount of money we requested was kept in trust by the loan officer and financially we were ready to go ahead with the construction. The institution didn't charge us any interest on the loan until we made our first withdrawal. Then they only charged us for the amount that was used. When I needed the money all I had to do was call the loan officer to tell him how much I needed and he issued it.

As it turned out I needed more than I figured I would for the interim financing. The officer was very understanding and let us sign new papers to cover the difference allowing us the rest of the financing we needed. Always pay by check as it is the one means of proving you paid a particular bill. When you get the financing they will inform you about a *lien waiver*.

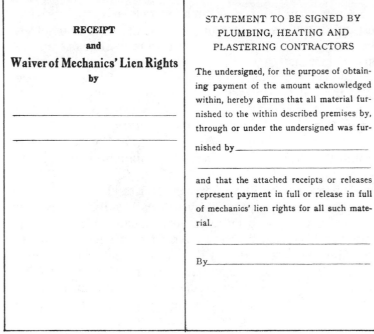

Fig. 10-5. Continued

LIEN WAIVER

A lien waiver, as shown in Fig. 10-5, is a form of receipt that you must get from every subcontractor or material supplier when you pay them. I can't stress enough the importance of the lien waiver. It should be filled out completely explaining exactly what is being paid for. This note protects you from anyone who would try to collect a second time. In many instances, it is the only legal note you have to prove you paid a bill, even though you may also have the cancelled check. For example, if a subcontractor gets paid from you and he doesn't pay a supplier for the material he used in your home or a laborer who worked on your home, they in turn could put a lien on your home to make you pay them for you are the overall contractor in charge. If you have a lien waiver they can't touch you and they are forced to go back to the subcontractor.

Before you start your home, check with the local and state governing bodies to find out what laws there are to protect the average citizen in this area. Remember, get the lien waiver signed at the time you hand over your check for payment.

FINAL PAYOFF AND SIGNING THE MORTGAGE PAPERS

When your home is completed return to the original institution you have the mortgage pledge with and inform them of your position. Tell them you're ready to close and set up your new mortgage.

To interject, about a month before you are ready to close, go to the county abstract office and apply for a copy of the abstract of your land. They will need a week or two to do a complete history and update of the property. Tell them when you will need it and give them ample time. You will have to submit the abstract to complete the remaining portion of the financing.

The institution will send an inspector to your new home to verify their requirements. They will call you in and remind you to bring in the balance of your unpaid bills. They will authorize the balance of the payments on checks from the institution. The loan officer will then call in the subcontractors to sign the lien waivers and receive the checks. Also, the officer will issue you a check to cover the interim financing from the second institution. When all the payoffs are made you will be credited the original commitment fee and charged the standard service fees the institution has for issuing a new mortgage.

There are various types of mortgages that financial institutions will issue. You must sit down to examine each one and select one you feel will be best suited to your needs.

If you are building on a piece of property that has never been fully taxed before, check around the area to find out what the average tax is. Count the number of calendar months you are going to occupy your new home during that tax year. If it is only four months, you can figure your first year's tax will be about 33—40% of what everyone else will pay that year. Inform your loan officer so that he has a figure to go by when he designs your payment. Your payment, to make it clearer for you, should include principle, interest, taxes, and insurance, although insurance may not be required.

If for some reason you have moved into your home and can't get the mortgage cleared because one or two subcontractors haven't finished their work there is no cause to be alarmed. In our case, the bank made all the payments possible, including the interim financing note, and charged us interest for the amount of the loan until the other subcontractors had completed their work. The interest they charged us was in lieu of any monthly payment we would have had to make for that period of time. When the remaining subcontractors were finished we completed the mortgage forms, the monthly payment schedule was set up, and the project was completed.

This was my first experience in dealing with financial institutions and it gave me some valued experience. As with all the other fields I came in contact with, it was extremely interesting, challenging, and something about which I would like to learn more.

TAXES

Every local government has their own tax structure. It would be extremely difficult for me to tell you what you will eventually end up paying. But here are a few guidelines that you might be able to follow.

It is one thing to be able to afford to build a house and it is another thing to be able to afford to pay the taxes on it.

In our case, I checked with the local governing body to find out what the people around our area were paying. They quoted me an assortment of figures because they varied with the front footage of the property, the size of house, what was on and in the house, and if there were any special assessments against the property.

To get our approximate tax figure, I gave the local tax office the amount of front footage our lot has on the street side, the amount of square footage the home would have, and the approximate building cost. From these figures they were able to estimate an approximate tax figure. But since the tax was also determined by what goes on and in the home there would be a variation. The local tax board gave me a schedule of what would add to the value of our home regardless of the base cost. The following list shows a sample of items that added to the tax base of our home, and by how much.

- Third ½-bath/powder room—$400
- Whole house air-conditioning—$1000
- Brick trim—$600
- Fireplace—$800
- Sliding patio door—$900
- Inside grill—$50
- Outside grill—$150
- Deck—$240

This schedule is not complete and it is dated. I am sure there are many items that I have missed and some that have increased in value, but these are what I was interested in at the time of building.

As you can see it is very easy to raise the value of your home just by adding a few items that the average home doesn't have. So your tax will depend on the value of your lot, the cost of the home you build, the assessments, if any, and the added values you build into your home.

CONCLUSION

Now that you have seen that subcontracting your home can be done, let's take one last look at where the savings come from.

- You will save the contractor's commission.
- As your own contractor you are entitled to a contractor's discount from most of the material suppliers.
- When you hire subcontractors they are hired by the job so you don't get involved in the direct payment of the laborers' fringe benefits that a professional contractor must pay.
- You may be able to buy a better piece of property than a particular contractor wants you to buy.
- Your subcontractors won't worry about using better quality material because you will determine the kind they use.

So you see, you can save in almost every area in the construction of your home. Since nothing is free except your time, by using it you can raise your housing standard a lot higher than your payments. You can now live like you've always wanted to and save 50% before you know it. Good luck!

11
Building Our Dream House

Up to now I have explained how we built our dream house and how you can build yours. The following section is an outline of the physical structure and a breakdown of what it took to build our home day by day, including the labor and material costs. Nothing is left out. It shows you what was done and the amount of time it took to do it. In the preceding pages all the information we gathered was used to construct the dream home we now live in. It took time—389 days, including all of the planning stages, but we had more time than money.

DAY-BY-DAY ACCOMPLISHMENTS

Day 1
Excavator digs foundation.
Inspections: foundation.

Days 2—6
Nothing.

Day 7
Cement men set footings and start constructing basement walls.

Days 8—10
Cement men complete basement walls.

Day 11
Waterproofer sprays outside of basement walls with hot asphalt to waterproof exterior.
Deliveries: first load of lumber for rough-in.
Inspections: before backfill.

Day 12

Owners hang polyethylene liner on outside of basement and pack dirt against the base to hold it in place. The liner acts as an additional water barrier.

Day 13

Nothing.

Day 14

Carpenters shore up interior of basement walls for added support against backfilling of dirt.

Deliveries: steel beams and columns.

Day 15

Excavator backfills dirt against basement. Carpenters set beams across basement walls and support columns under the beams. Basement walls completely capped.

Days 16—18

Carpenters nail subflooring in place. First-floor interior and exterior walls are built and installed.

Deliveries: second load of lumber for rough-in.

Days 19—20

Nothing.

Day 21

Carpenters nail in place the second-floor subfloor. Exterior walls built and installed.

Day 22

Carpenters complete and install second-floor interior walls (See Fig. 11-1).

Days 23—25

Carpenters rough-in and enclose entire roof. Owners clean first and second floor.

Deliveries: third load of lumber for rough-in on day 23; fourth load of lumber for rough-in on day 25.

Inspections: rough-in.

Day 26

Nothing.

Day 27

Carpenters start to shingle roof. Plumbers start to rough-in pipes.

Fig. 11-1. Frame-up and wall construction is completed. The house is now ready for roofing by carpenters.

Deliveries: rough-in outlets (for light fixtures), shingles, aluminum siding, tar paper, nails, backerboard, and flashing paper.

Day 28

Electricians start to rough-in their boxes and conduit. Heating men inspect rough-in area and install most of the ductwork. Telephone man roughs-in outlets on all floors. Plumbers and carpenters continue their roughing-in. Owners clean up after subcontractors.

Deliveries: outside doors, windows, and patio door.

Day 29

Plumbers finish rough-in and install bathtubs. Electricians continue to rough-in. Brick men install 75% of stack. Carpenters continue to shingle roof.

Inspections: plumbing rough-in.

Day 30

Carpenters finish shingling roof, rough-in kitchen soffit, cut out opening for fireplace, and start to install windows.

Day 31

Carpenter finishes installing windows, patio door, and exterior doors. Excavator fills in and levels dirt floor in garage. Electrician continues to rough-in conduit.

Day 32

Electrician finishes the electrical rough-in.

Day 33

Owners clean up inside and outside of house.

Day 34

Inspections: electrical rough-in.

Day 35

Nothing.

Day 36

Electricians return to adjust conduit to help the heating men reposition one run of their ductwork.

Day 37

Heating men finish installing ductwork.

Day 38

Inspections: heating/cooling rough-in.

Day 39

Nothing.

Day 40

Owners clean up inside of house.

Day 41

Brick man checks exterior of house to make ready for exterior brick facing.

Deliveries: finishing bricks and cement.

Day 42

Heating man installs clothesdrop. Brick man starts building brick front and exterior of fireplace. Owners clean up inside of house.

Day 43

Brick men finish exterior facing and outside and inside of fireplace. Carpenters start to put siding on exterior. (See Fig. 11-2.)

Deliveries: aluminum siding accessories.

Day 44

Carpenters continue to install siding.

Deliveries: insulation, polyethylene liner, and firewall Sheetrock.

Returns: lumber not used at rough-in.

Day 45

Brick man returns to clean facing brick. Carpenters continue installing siding.

Fig. 11-2. Brickwork is completed on front of the house. Foil-faced insulation is installed to make exterior ready for aluminum siding.

Day 46

Carpenters switch to installing firewall in garage.

Day 47

Nothing.

Day 48

Carpenters finish installing firewall in garage and return to installing exterior siding.

Day 49

Nothing.

Day 50

Carpenters start to insulate the house.

Day 51

Carpenters finish installing insulation and polyethylene water barrier and begin to hang plasterboard.

Deliveries: plasterboard, nails, tape, and joint topping.

Day 52

Carpenters finish installing exterior siding, except for soffit.

Day 53

Carpenters return to cut, fit, and hang plasterboard.

Day 54

Owners clean up main floor and exterior.

Day 55

Carpenters continue to hang plasterboard. Plumbers start to dig in front yard to install connecting sewer pipe.

Day 56

Carpenters finish downstairs plasterboard and start on second floor. Plumbers finish sewer connection and replace dirt.

Inspections: sewer system.

Day 57

Carpenters finish installing plasterboard on second floor and start to cover exterior soffit with siding. Owners clean upstairs.

Day 58

One carpenter starts to tape the seams and cover the nails in the plasterboard on the second floor. Other carpenters continue to cover the soffit with siding and hang the garage door. Owners clean up second floor.

Deliveries: garage door.

Day 59

Carpenters put second and third coat of plaster over seaming tape on second floor walls then start on first floor. No work done on soffit. Owners clean second floor and part of first floor.

Day 60

Carpenters finish soffit. Owner digs hole and buries drain barrel in garage floor.

Day 61

Nothing.

Day 62

Carpenters hang screen doors and continue to tape and cement walls on first floor.

Day 63

Carpenter puts second coat of cement over seaming tape and outside corners, sands and completes small second floor bathroom and part of second floor. Cement men level basement stairs and spread sand over dirt basement floor to level in preparation for pouring cement floor.

Deliveries: finished flooring and rosin paper.

Day 64

Owners clean up first and second floors.

Day 65

Cement men pour entire basement floor, set up side support forms for front and back steps, and lay wire on dirt floor in garage. Owners clean up outside and first floor.

Deliveries: cement.

Day 66

Cement men pour entire garage floor, apron, and front and back steps. Carpenters finish sanding the second floor plasterboard walls and apply the third coat of cement over the seams on the first-floor walls. Owners clean up second-floor bathrooms. Owner paints base coat in second floor bathrooms.

Deliveries: cement.

Day 67

Carpenters finish last coat over seams on first floor, then start laying rosin paper and versaboard over subfloor on second floor. Owner paints second coat in two bathrooms and cleans up first floor.

Day 68

Owner cleans up outside of home.

Day 69

Carpenter sands first-floor walls and completes nailing versaboard flooring over subfloor on the second floor.

Deliveries: interior doors, window and door casings, wall baseboard, window and door accessories.

Day 70

Carpenters cover second-floor finished flooring with plastic to protect against falling plaster when ceiling is cement-sprayed.

Day 71

Carpenters finish sanding first-floor walls to prepare for spraying cement on ceiling.

Deliveries: kitchen cabinets, vanities, and countertops.

Day 72

Plasterers spray cement on ceiling and add sparkles to one bedroom, living room, dining room, and family room.

Day 73

Telephone man installs jacks in master bedroom, kitchen, family room, and office. Stainer puts first coat on vanities, trim, and moldings. Carpenters lay rosin paper over subfloor and nail versaboard on as finished floor in areas later to be covered by carpeting, and plywood flooring in areas later to be covered by linoleum.

Day 74

Carpenters sand and touchup walls on both floors; install paneling in family room, dining room, part of foyer, and stairway; and install inside window trim. Owner paints second coat in master bedroom, boy's bedroom, and part of mud room.

Deliveries: vanity bowl tops.

Day 75

Owner paints first coat in kitchen work area. Owners clean up outside area around house.

Day 76

Electricians wire house, place recepticals in basement and outside for lights, and install bathroom vent fans. Carpenters install sliding pocket door in family room, finish paneling family room, and hang clothespoles in closet. Plumber hooks up water to home.

Deliveries: paneling and moldings, electrical fixtures and bulbs, doorbell and chime.

Returns: unused Sheetrock and flooring.

Day 77

Carpenters finish paneling foyer, stairway, upstairs hall, first-floor bathroom, and kitchen eating area and hang beams in family room.

Deliveries: Vent-A-Hood, Char-glo-grill, and ductwork.

Day 78

Carpenters finish installing inside window trim on remaining windows, hang remaining internal doors, put unstained molding on base of walls, and install permanent stairtreads.

Day 79

Carpenters install complete aluminum facia on outside of house and aluminum exterior trim for base of windows.

Deliveries: chopping block, trim moldings, and dishwasher.

Day 80

Heating men bring in and install main furnace. Owner cuts paneling for childrens' bedrooms and finishes painting mud room.

Day 81

Owner finishes installing paneling in girl's room.

Day 82

Owner finishes installing paneling in boy's room and clean up in both girl's and boy's rooms. (See Fig. 11-3.)

Day 83

Electricians finish hooking up fixtures in outside area, basement, kitchen, girl's bedroom, and upstairs hallway. Stainer finishes first coat of all trim, doors, and molding except kitchen cabinets. Heating men begin installing ductwork in the basement to hookup furnace.

Day 84

Stainer completes first coat of kitchen cabinets and puts sealer coat on rest of woodwork. Heating men continue to run

Fig. 11-3. Boy's bedroom is finished with paneling, cork walls, and sliding closet doors.

ductwork in the basement. Owner installs trim molding in girl's bedroom and touches up stained areas of walls in childrens' rooms.

Day 85

Heating men finish all basement ductwork and complete hookup to large furnace. Stainer finishes master bedroom trim, bathroom vanities, and doors on second floor; sands down windows and trim; and finishes sealing trim and molding. Owner puts on second coat of paint in childrens' bedrooms and covers stain on walls in master bedroom.

Day 86

Stainer finishes all kitchen cabinets, clothesdrop door, exterior doors, and closet bifold doors. Carpenters complete exterior aluminum window-sill trim, hang closet bifold doors, put on door handles, trim kitchen cutting board, and start to put on molding in second floor hallway. Heating men check out installed ductwork, cold- and hot-air registers, hook up piping to exterior gas meter, and install small furnace. Ceramic tile man sets tiles in both second-floor bathrooms. Owner cleans up both floors.

Deliveries: bathroom mirrors.

Day 87

Carpenters finish the moldings in the upstairs hallway, put baseboard molding in all three bedrooms, hang remaining closet door, cut out space for kitchen sink in Formica countertop, install countertop, and set sink in place for hookup. Heating man ducts out small furnace for basement heating. Owner paints second coat in master bath and touches up boy's and girl's bedrooms (see Fig. 11-4). Ceramic tile man applies grout between all tiles and cleans them. Linoleum man installs linoleum in foyer, mud room, first-floor bathroom, and large bathroom on second floor.

Inspections: final heating.

Day 88

Carpenters hang bathroom mirrors, put handles on kitchen cabinet doors and bathroom vanities, set Vent-A-Hood in place on wall, cut chopping block, and set in gas grill. Both block and grill placed on top of designated kitchen cabinet under Vent-A-Hood. Carpenters also finish installing baseboard moldings on first floor and finish ceiling trim in family room.

Deliveries: four sets of bifold closet doors.

Fig. 11-4. Girl's bedroom is finished with painted walls and sliding closet doors.

Day 89

Owner paints mud room and cleans up inside and outside the house.

Day 90

Stainer stains and varnishes last four sets of bifold closet doors, fills putty in nail holes, stains extra baseboard, and touches up woodwork. Plumber installs finishing pipe, installs and hooks up laundry tubs in basement, installs hot water tank and makes it ready for hookup. Owner trims wood with masking tape in kitchen and ceramic tile in second floor bathrooms to prepare for final coat of semigloss paint.

Day 91

Plumber installs all toilets and sinks and hooks up kitchen sink.

Day 92

Plumber finishes all odds and ends of gas-piping including connecting range and grill, hooking up garbage disposal, installing water pipe for dishwasher and clotheswasher installing and vent for dryer. Owner paints final coat in mud room and masks living room trim to make ready for painting.

Inspections: final plumbing.

Day 93

Plasterers respray ceilings in living room and dining room.

Day 94

Bricklayer touches up fireplace, seams wall where plasterboard and fireplace meet, cements clothesdryer vent in place, and cements outside brick facing where facing and aluminum siding meet. Owner applies first coat of paint to living room walls and covers stain marks on kitchen walls and ceiling to prepare for second coat.

Day 95

Carpenter finishes all but one doorway molding, including lattice work in dining room, hallway, handrailing, trim in kitchen, molding for patio door, and trapdoor for inside living room ceiling, and enclosed clothesdrop. Owner paints second coat in kitchen and living room.

Day 96

Owner paints third coat in kitchen and second-floor bathrooms and cleans areas where painting was completed.

Day 97

Owner paints trim in girl's room to match color of paneling.

Deliveries: carpets.

Day 98

Stainer fills putty holes and checks for areas that require touchup. Owner puts second coat of paint on trim for girl's room, paints several registers and cold-air returns to match background, and cleans up kitchen area.

Day 99

Owner varnishes trim for girl's room, puts second coat on some registers and first coat on others, reinstalls second-floor registers, and puts cork tile on wall in girl's room. Owners clean up remaining mess in bathrooms upstairs.

Day 100

Ceramic tile man finishes ceramic floor molding; installs towel, paper, soap, and toothbrush holders; and touches up areas missed the first time. Owner installs cork tile on wall in boy's room, installs air returns, cleans bedroom and bath, and makes master bedroom and bath ready for carpet. Stainer returns to stain extra moldings for future use.

Deliveries: air-conditioner.

Day 101

Electricians hook up remaining light fixtures, dishwasher, and garbage disposal; wire furnace for air-conditioner; set up and install thermostat in dining room and basement; set up circuit breakers for all rooms; and complete electrical wiring for final inspection. Excavator backfills and levels back and front yards. He brings in five loads of sand and two loads of dirt for front yard. Owners clean up stairway, hallway, living room, dining room, and family room floors to prepare for carpet and professionals to install it in those areas the next day.

Inspection: final electrical.

Day 102

Carpet layers bring and install carpet in the living room, dining room, family room, stairway, and hallway upstairs. Owner starts to lay carpet in master bathroom and sets carpet in master bedroom.

Deliveries: carpet.

Day 103

Owner finishes laying carpet in master bathroom and sets carpet in master bedroom.

Day 104

City gasmen start to hookup gasline on outside of house. Landscaper regrades driveway to prepare for blacktop and plots endpoints for blacktop court in backyard. Carpenter finishes trim in boy's and girl's room and foyer. Owners finish laying carpet in master bedroom and line kitchen cabinets.

Day 105

Landscapers finish grading driveway, lay and level blacktop, and regrade part of backyard to prepare for blacktop court. Heating men hook up air-conditioner to furnace. City gasmen complete gas hookup. Owner touches up trim in girl's and boy's rooms, lays carpet in girl's room, and prepares boy's room for carpet.

Day 106

City gasmen turn meter on and check gas flow. Electrician checks to make sure house has full electric power and air-conditioning is in proper working order. Landscapers lay and finish blacktop court. Owners finish laying carpet in girl's room, set carpet in boy's room, and install shelves in kitchen canned goods closet.

Day 107

Owners finish laying carpet in boy's room, prepare carpet in kitchen for installation, do final cleaning of bathrooms, line closet shelves, and move in clothing.

Day 108

Owner finishes laying kitchen carpet and does last-minute preparations for moving in the next day.

Day 109

Owners move in with furniture, clothes, children, large mess, and more confusion.

Day 110

Owner hooks up gas range, grill, dryer, and washer.

Day 111

Landscapers level out front and back yards and lay sod in back third of yard.

Inspection: final completion.

Day 112

Landscapers fix low area in blacktop court and finish sodding back yard.

Day 113

Landscapers finish sodding side and front of home.

Day 114

Owners rest, relax, and enjoy new home.

FINAL BREAKDOWN OF LABOR COSTS

The following is a list of the subcontractors and the amount of man-hours it took to complete the various rough-in and final stages of their work. When comparing the following schedule, you must keep in mind the number of hours worked in proportion to the size of the home being constructed. The figures below are based on a home with an actual living area of 1250 sq ft in the basement, 1400 sq ft on the first floor, 850 sq ft on the second floor, and a two-car attached garage.

The chart (Fig. 11-5) is computed in man-hours and in most instances two or more men worked together. All of their hours are combined in the totals.

SUBCONTRACTOR	PART OF HOME CONSTRUCTION	NUMBER OF HOURS WORKED	LABOR COST
Excavator	Dig foundation	9	
	Backfill	5	
	Level garage floor	3	
	Level front and back yards	8	
	Total	25	$325
Brick & concrete subcontractor	Build basement walls for house and garage	108	$1200
	Construct brick front and fireplace	45.5	$850
	Lay basement and garage floor plus front and back steps	93	$1000
	Total	246.5	$3050
Plumbers	Rough-in	38	
	Finish	20	
	Total	58	$1400
Heating & cooling subcontractor	Rough-in	23.5	
	Finish	60	
	Total	83.5	$1800
Electrician	Rough-in	62	
	Finish	58	
	Total	120	$1100
Ceramic tile & linoleum subcontractor	Ceramic tile installed	18.5	
	Linoleum installed	8	
	Total	26.5	$300
Stainer	Finish	63	$485
Plumber	Sewer hookup	16	$375
Asphalt & sod subcontractor	Driveway	27	
	Blacktop court	42	$500
	Sod	46	$375
	Total	115	$875
Garage door installer	Install garage door	4	$30
Carpet installer	Install carpet	10	$165
Telephone Co	Rough-in	1	n/c
	Install four jacks	2.5	$55
	Total	3.5	$55
Carpenter	Rough-in	200	
	Shingle roof	82	
	Install windows and outside doors	42	
	All exterior siding, trim, moldings, soffit, and facia	193	
	Garage firewall	24	
	Insulation and polyethylene wall liner	34	
	Install finish floor	33	
	Hang plasterboard	42	
	Finish plasterboard	38	
	Spray ceilings	11	
	Install paneling	53	
	Finish trim, hang doors, doorknobs, install cabinets and vanities, etc.	56	
	Total	907	$6150
City gas dept	Gas hookup	26	n/c
Owner	Install carpet	28.5	n/c
Owner	Painting	88	n/c
Owner	Cleanup	101	n/c
	Grand Total	1921.5	$16,110

Fig. 11-5. Breakdown of labor costs.

Upon looking at the number of dollars spent for the man-hours worked, you must take into consideration that you are getting an expert in each field. You are buying the man, his technical knowledge, the equipment he must have to do the job, a warranty of his work for a given length of time, and the importance of your job to the success of his business.

In the long run, the price you pay will depend on the quality of work your subcontractor will produce.

FINAL BREAKDOWN OF MATERIAL COSTS

The following list (Fig. 11-6) covers the cost of materials purchased for both the rough-in and final stages of building our home. They were purchased through distributors and subcontractors.

If you live in a state that has a state sales tax, check to see what percentage of your total materials charge went to pay that tax. This may be deductible from your state income tax.

SUPPLIER	MATERIAL	COST
Excavator	Fill sand and dirt	$125
Brick and concrete subcontractor	Concrete blocks, facing brick, and cement	$3,200
Plumber	All materials (except vanity tops)	$1,350
Plumbing supplier	Vanity tops	$175
Heating/cooling subcontractor	Heating plants, air-conditioner, ductwork, air cleaner, humidifier	$1,500
Electrician	All rough-in equipment including switches and outlets	$575
Electrical supplier	Fixtures	$675
Ceramic tile and linoleum subcontractor	Ceramic tile and linoleum	$400
Asphalt and sod subcontractor	Asphalt and sod	$700
Garage door subcontractor	Garage door	$140
Carpet companies	Carpets	$2,000
Paint store	Paint and mirrors	$175
Steel company	Steel beams and columns	$320
Lumber company	Lumber	$7,450
Millwork company	Windows, doors, trim	$3,250
Lumber company	Kitchen cabinets	$1125
Lumber company	Kitchen countertops	$155
Roofing and siding subcontractor	Roofing material	$540
Roofing and siding subcontractor	Siding material	$1400
Appliance store	Dishwasher	$200
Waterproofing subcontractor	Asphalt	$85
Vent-A-Hood company	Grill and Vent-A-Hood	$325
Department store	Window shades	$75
Drapery store	Draperies and rods	$800
	Grand Total	$26,740

*Subtotals and grand total include 4% State Sales Tax

Fig. 11-6. Breakdown of material costs.

CATEGORY	DOLLAR COST	ACCUMULATIVE DOLLAR COST	SQ FT COST	ACCUMULATIVE SQ FT TOTALS
Basic home, basement & two-car garage	$38.300	$38.300	$17.00	$17.00
Carpet and pad	$2150	$40.450	$01.00	$18.00
Sod and blacktop	$1600	$42.050	$00.75	$18.75
Rods and draperies	$800	$42.850	$00.35	$19.00
Property	$9500	$52.350	$04.25	$23.25
Financial charges	$500	$52.850	$00.25	$23.50

Fig. 11-7. Actual square-foot cost of the completed house.

MATERIALS FROM SUBCONTRACTORS

The following subcontractors ordered and scheduled their own material deliveries as included in their bids.

- Brick and concrete company.
- Plumbing company.
- Electric company (Everything but lighting fixtures.)
- Heating/cooling company.
- Linoleum and ceramic tile company.

ACTUAL COST OF THE HOME

The following chart (Fig. 11-7) is a breakdown of the total cost, rounded off, for the 2250 sq ft above-ground living space of our home broken down into five categories.

If these figures seem like a great deal of money you're absolutely correct, but they should be compared to what some custom, semicustom, and precut builders would charge. Since the land you build on is a separate figure, depending on its size, location, and cost, just consider what these other builders would charge for the equivalant package. In our area, the cost to construct the home we built, using the same quality materials, by a contractor is between $40 and $45 per square foot, about twice as much as we paid by subcontracting our own construction.

As mentioned several times throughout this book there are many variables, depending on where you build and what you build, that are all part of the puzzle you must piece together. Search these variables out. Check them and use them as guidelines to understanding what you can and cannot build.

BLUEPRINTS AND MAN-HOUR CHARTS

Following are the blueprints (Figs. 11-8—10) for our dream house and a day-by-day chart of the man-hours (Fig. 11-11) worked by each subcontractor.

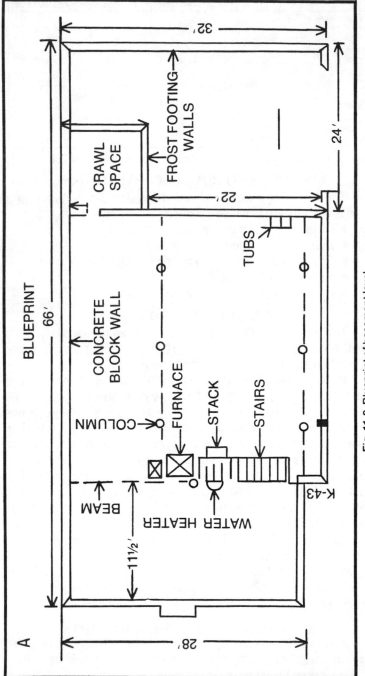

Fig. 11-8. Blueprint of basement level.

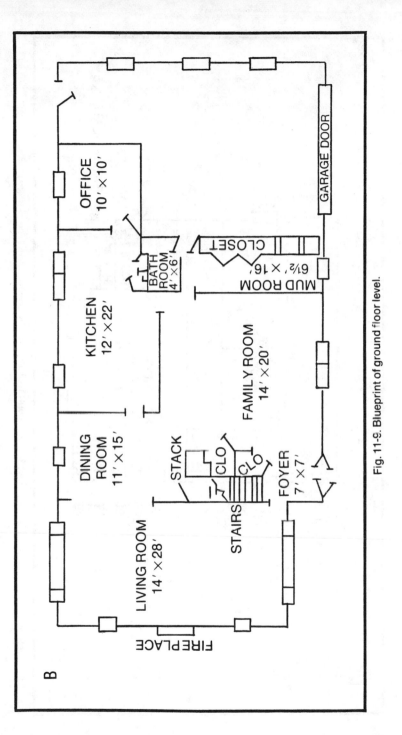

Fig. 11-9. Blueprint of ground floor level.

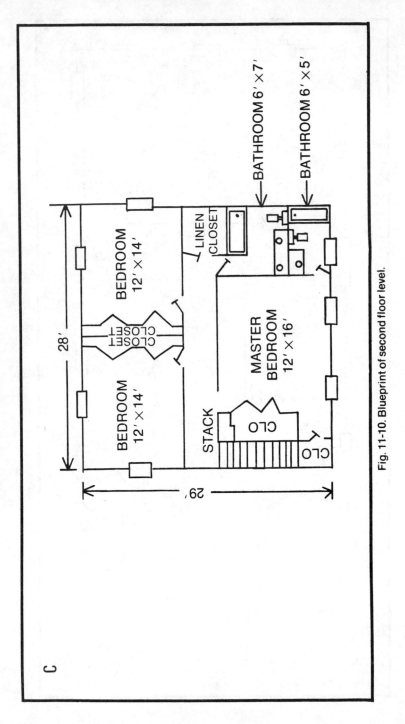

Fig. 11-10. Blueprint of second floor level.

BEDROOM 12' × 14'

CLOSET
CLOSET

BEDROOM 12' × 14'

LINEN CLOSET

MASTER BEDROOM 12' × 16'

STACK

CLO

CLO

BATHROOM 6' × 7'

BATHROOM 6' × 5'

28'

29'

C

DAY	EXCAVATOR	BRICK LAYERS	CARPENTERS	CLEANUP	INSTALL POLYETHYLENE	ASPHALT
1	2 men 4½ hr ea					
2						
3						
4						
5						
6						
7		2 men 4 hr ea				
8		4 men 10 hr ea				
9		4 men 10 hr ea				
10		4 men 5 hr ea				
11						1 man 1 hr
12					2 men 2½ hr ea	
13						
14			2 men 1½ hr ea			
15	1 man 5 hr		4 men 9 hr ea			
16			4 men 9 hr ea			
17			4 men 9 hr ea			
18			3 men 1 man 9 hr ea 4½ hr			
19						
20						
21			3 men 9 hr ea			
22			3 men 9 hr ea			
23			4 men 9 hr ea			
24			4 men 9 hr ea			
25			3 men 9 hr ea	2 men 3½ hr ea		

Fig. 11-11. Man-hour chart, day 1 through 25 (page 1 of 5).

DAY	EXCAVATOR	BRICK LAYERS	CARPENTERS	PLUMBING	HEATING	ELECTRICAL	CLEANUP
26							
27			2 men 9 hr ea	2 men 8 hr ea			
28			3 men 9 hr ea	2 men 8 hr ea	1 man 8 hr 1 man 4 hr	2 men 8 hr ea	
29		2 men 3 hr ea	3 men 9 hr ea	2 men 2½ hr ea		2 men 8 hr ea	
30			2 men 9 hr ea 1 man 4 hr				
31	1 man 3 hr		1 man 9 hr			2 men 4 hr ea	
32						3 men 6 hr ea	
33							2 men 2½ hr ea
34							
35							
36						2 men 2 hr ea	
37					3 men 3 hr ea		
38							
39							1 man 2 hr
40							
41		1 man 1 hr					
42		4 men 8 hr ea			1 man 2½ hr		2 men 2½ hr ea
43		4 men 8 hr ea	3 men 7½ hr ea				
44			2 men 8½ hr ea				
45		1 man 3 hr	2 men 8½ hr ea				
46			2 men 8 hr ea				
47							
48			2 men 8½ hr ea				
49							
50			2 men 8½ hr ea				

Fig. 11-11 (cont). Man-hour chart, day 26 through day 50 (page 2 of 5).

DAY	EXCAVATOR	BRICKLAYERS	CARPENTERS	CLEANUP	SEWER HOOKUP	PAINTING	SPRAY CEILING
51			2 men 8½ hr ea				
52			2 men 8½ hr ea				
53			2 men 8 hr ea				
54				2 men 3½ hr ea			
55			2 men 8½ hr ea		2 men 4 hr ea		
56			2 men 8½ hr ea		2 men 4 hr ea		
57			2 men 8½ hr ea	2 men 3½ hr ea			
58			1 man 10 hr 1 man 3 hr 2 men 8½ hr ea	1 man 3 hr			
59			1 man 9 hr 1 man 4 hr	1 man 3 hr			
60	1 man 2 hr		2 men 8½ hr ea	2 men 2 hr ea			
61							
62			2 men 8½ hr ea				
63		1 man 8 hr	1 man 9 hr				
64				1 man 1½ hr			
65		3 men 9 hr ea 1 man 8 hr 1 man 5 hr		2 men 3 hr ea			
66		1 man 5 hr 3 men 8 hr ea	2 men 8½ hr ea	1 man 2 hr		1 man 4½ hr	
67		2 men 8½ hr ea		1 man 2 hr		1 man 4 hr	
68				2 men 2½ hr ea			
69			1 man 8 hr				
70			1 man 2 hr				
71			2 men 2 hr ea				
72							2 men 4 hr ea
73			2 men 8 hr ea 1 man 11 hr			1 man 4 hr	
74			4 men 8½ hr ea			2 men 4 hr ea	
75				2 men 2½ hr ea		1 man 3 hr	

Fig. 11-11 (cont). Man-hour chart, day 51 through day 75 (page 3 of 5).

DAY	BRICK LAYER	CARPENTERS	PLUMBING	HEATING	ELECTRICAL	PAINTING	STAINING	CLEANUP	CERAMIC TILE	LINOLEUM
76		1 man 8½ hr	1 man 1 hr		2 men 8 hr ea					
77		2 men 9 hr ea								
78		2 men 8½ hr ea								
79		2 men 8½ hr ea								
80		1 man 1½ hr		2 men 1½ hr ea						
81		1 man 4½ hr								
82		1 man 5½ hr								
83				2 men 1 hr ea	2 men 8 hr ea		1 man 12 hr			
84				2 men 6 hr ea		2 men 4 hr ea	1 man 11 hr			
85				2 men 8 hr ea		2 men 4 hr ea	1 man 12 hr			
86		2 men 8½ hr ea		2 men 8 hr ea 1 man 3 hr			1 man 10 hr	1 man 1½ hr	1 man 8 hr	
87		1 man 8 hr		1 man 2 hr		1 man 3 hr			1 man 8 hr	1 man 8 hr
88		2 men 8 hr ea								
89						1 man 3 hr		2 men 3 hr ea		
90			1 man 6 hr			1 man 2 hr	1 man 3 hr			
91		1 man 8 hr								
92		1 man 6 hr			2 men 2½ hr ea					
93		1 man 3 hr								
94	1 man 1½ hr					1 man 4½ hr				
95		1 man 7 hr				2 men 4½ hr ea				
96						1 man 5½ hr		2 men 3 hr ea		
97						1 man 1½ hrs				
98						1 man 4½ hr	1 man 3 hr	1 man 1 hr		
99						1 man 8½ hr		1 man 1 hr		
100						1 man 2 hr	1 man 3 hr	2 men 4 hr ea	1 man 2½ hr	

Fig. 11-11 (cont). Man-hour chart, day 76 through day 100 (page 4 of 5).

DAY	EXCAVATOR	CARPENTERS	HEATING	ELECTRICAL	CLEANUP	CARPET LAYING	ASPHALT	GAS HOOKUP	MOVING	SOD
101	2 men 4 hr ea			1 man 4 hr 1 man 8 hr 1 man 10 hr	1 man 4 hr					
102						2 men 5 hr ea 1 man 1 hr				
103		1 man 2 hr 1 man 1½ hr				1 man 7½ hr	2 men 3 hr 1 man 2 hr	3 men 5 hr ea		
104		1 man 1½ hr					1 man 7 hr 4 men 5 hr ea	3 men 3 hr ea		
105			2 men 3 hr ea		1 man ½ hr		4 men 8 hr ea			
106				1 man 4 hr	1 man 3 hr	1 man 4 hr	1 man 2 hr			
107					1 man 4 hr	1 man 8 hr				
108					1 man 3 hr	1 man 6 hr				
109									3 men 9 hr 1 man 8 hr	
110									2 men 1½ hr ea	
111										1 man 5 hr 3 men 3½ hr ea
112										4 men 1 hr ea
113										2 men 4 hr ea 3 men 6 hr ea

Fig. 11-11. Man-hour chart, day 101 through day 113 (page 5 of 5).

12

Helpful Hints

Since every professional has several tips and hints stockpiled through years of experience to help him do his work faster and easier, I gathered what I thought might be a few of the most important hints and grouped them under the subcontractor's category to which they belong. Other helpful tidbits have been sprinkled throughout the book.

BLACKTOP
- Don't be afraid to lay blacktop before it rains. The rain will act as a cooling agent and speed the hardening process of the blacktop.
- Stay off blacktop for five to seven days after it is poured to make sure it is firm.

CARPENTRY
- Whenever possible use oversized doorjambs and half-inch plasterboard for walls that will be paneled. This will give you a much stronger wall.
- Before taping and seaming plasterboard walls clean the subfloors. Tape some inexpensive plastic covering over the floor to catch wet plaster that is spilled or dust from sanding during the finishing phase. This can easily save you up to ten hours of cleaning and scraping time.
- Put a layer of red or gray rosin paper between the subfloor and top floor. It will absorb moisture, prevent floor squeaks, and keep dust to a minimum.

CERAMIC TILE
- After ceramic tile is installed spray the grout between the tiles with silicon. This will act as a coating agent and prevent the grout from yellowing.

- If a vanity is installed in an area where ceramic tile will be placed adjacent to it, make sure that the vanity is stained and varnished first. If any grout is spilled or splashed on raw wood it will work its way into the wood and discolor any future staining.

HEATING
- Always put insulation behind ductwork that is placed in outside walls. Have your heating/cooling subcontractor install his rough-in ductwork before the electricians or plumbers install their equipment. It is easier for other subcontractors to work around ductwork than it is to run ductwork around other equipment.
- When heating a home with a split entry or double doors always put a hot-air duct in the entry. Be sure it can be angled at the doorway to stop cold air from coming into the house in the winter.
- Use six-inch ductwork where possible to heat a home. You will get about 20% more heat than from conventional five-inch ductwork. Also, your furnace won't have to work as hard to heat your home.

PLUMBING
- If you are going to install a large gas-powered water heater, range, dryer, etc., make sure the gasline entering the house is at least $1\frac{1}{2}$ in. in diameter. Standard gaslines are smaller and you may lose a pilot light or two if you are using too many appliances at one time.
- When hooking up gas appliances don't forget to bleed the line of excess air for three to five minutes before attempting to light the pilot.
- Before coupling the connecting lines from the appliances to the gasline, ring the threads of the gasline with a small amount of Teflon tape. This acts as a fantastic sealant to prevent leaks.
- Before you have the concrete floor poured in the basement, check with your plumber to make sure that he has the plumbing positioned properly.

STAINING AND PAINTING
- Check on the procedure that the stainer follows to complete his job. He should follow these steps: stain, seal, putty, sand, varnish.

- Hang all kitchen cabinets after they are stained. Each piece of wood is a different natural color. The stainer can match the shade better if he has all the woodwork in front of him at the same time.
- When cleaning the sealing varnish from the windows, leave a $1/_{16}$ inch rim around the edge of the glass. This seals the wood and prevents it from rotting since moisture cannot enter between the frame and the glass.
- Always use latex paint as the first coat over raw plasterboard. An oilbase paint applied directly to raw plasterboard may cause bubbling or shrinking of the outer paper liner.
- Before applying ceramic tile and adhesive to plasterboard walls, apply one coat of latex paint followed by a coat of semigloss oilbase paint over the plasterboard area. The adhesive holding the tiles in place will adhere better to the paint-covered wall.

13

Layman's Dictionary

This section includes the basic terms and definitions you will need to know to participate in a discussion with subcontractors and suppliers. You will have the information to understand how the subcontractors are able to blend their areas of work into a home for you.

Adjustable Elbows (90°). Used as an extension to turn a run around a corner or bend it to a desired shape.

Aluminum Channel. The track inside the window unit that allows them to slide up or down. There is more heat lost using an aluminum channel than a vinyl channel.

Aluminum(foil face). Available in widths of 15 to 23 in. and thicknesses of 2¾ to 7 in. It is placed between the exterior wall studs or the ceiling joists and is stapled to stay in place. Aluminum covers the outside exposed area.

Aluminum-Foil Vapor Barrier. Similar to the polyethylene sheeting in purpose, the outside cover is made of aluminum and acts as an interior vapor control over the insulation.

Aluminum Siding. It is available in a variety of thicknesses and colors in a smooth or embossed finish. Available in horizontal and vertical lengths, it can be mixed or matched with any other siding. Aluminum can be cut with a power tool.

Aluminum Trim Nails. They are used when putting aluminum siding and trim in place on the outside as they will not rust.

Amperes (amps). A measurement of electrical current. Most popular service boxes are sold with 100A, 150A, or 200A service. The higher the amperage, the more circuits are available for use. In an average home, a 150A service should be more than enough to service all the outlets.

Area Wall. This curved metal sheet is placed around the outside of the basement window. When the excavator regrades the property for the last time a pocket of space will be left around the basement window.

Asphalt Roofing Paper.

15 lb. This roofing paper should cover the entire roof area and lie as a base for the shingles to be nailed over. Fifteen lb equals the weight of the asphalt paper laid over a 10 × 10 ft area of the roof.

60 lb. This heavy paper should be used as an additional layer covering the first three ft measuring up from the outside edge of the roof. Sixty lb equals the weight of the asphalt paper laid out over a 10 × 10 ft area of the roof.

50 lb. This paper should be used as a second layer covering the second three feet up from the outside edge of the roof. Fifty lb equals the weight of the alphalt paper laid over a 10 × 10 ft area of the roof.

Note: The two heavier second layers of roofing paper act as double protection against a leaking roof, especially in the northern climates where snow accumulation and ice backup from frozen gutters are a problem.

Asphalt Roofing Shingles. They are available in a variety of colors, styles, and sizes to match the design of the home. The average structure can handle a 235 to 250 lb test shingle. Shingles with a test weight of more may require a roof with added support.

Assets. Any property or cash value one has that can be used to pay off debts.

Awning Window Unit. The unit has a window that works on a lever-type lock that pushes out and upward. The screen is attached to the inside of the unit.

Backsplash. It is molded into a single-unit vanity tops, but is a separate piece for the larger two-bowl vanity top.

Backwater Valve. It's attached to the underground pipes that lead to a sewer and is installed during rough-in. It contains a trapdoor or flap that permits waste to flow out freely; if the sewer should back up for some reason, the trapdoor or flap is forced shut by the pressure of the sewer water.

Barn Sash (stationary sash). A permanent, nonopening window.

Barn Siding. An eight-inch wide kiln-dried horizontal siding. The siding overlaps and is interlocking to make the exterior airtight.

Baseboard Register. An angled warm-air register that fits onto the baseboard area of the wall and functions the same as a floor register.

Base Cabinets. These cabinets stand on the floor and support the countertop. They are available in one and two door models.

Base Open-End Shelve. This unit can stand by itself or be attached to an exposed end of a cabinet. It's available in a half-circle or oval shape. Also known as what not shelves.

Basement Ceiling Angle Boot. Used with circular venting. Connecting to the venting, it has an opening to accept a register cover. Two flat pieces of metal attached to the sides of the opening are nailed to the studs to hold it in place.

Basement Window Unit. This window fits into a concrete block opening left by the cement men. Available in three sizes with a standard 27 in. width.

Base Vanity. A wood or metal storage cabinet that is positioned under the sink. It contains a storage area. It's available in a variety of widths with two opening doors. Also, some models are available with drawers.

Bathtub. The standard model is made of porcelain over steel or enamel over cast iron and is five ft long. When buying the tub check and order a right or left end drain. There are many varieties including recessed or sunken tubs. In addition, there are models made over five ft long.

Batten. Usually 1 × 3 in. strips of wood that act as a barrier and a decorative touch where two boards join. It also makes the exterior airtight.

Bifold Closet Doors (with frames). The set includes four doors. Two each are hinged together. The doors are prefitted and hang from an overhead track. They come in wood or metal louvered or plain face. They open from the middle toward the outside.

Blacktop. There are a variety of asphalt grades available. The best contain a powdered concrete mixture to keep the blacktop hard in warm weather. Finely crushed rock may be poured over the flat surface and rolled into it for additional strength and support. The blacktop should be poured over a compressed crushed-rock base.

Blueprint. A technical drawing, designed by an architect or draftsman, detailing the house. It covers all the technical aspects the subcontractors will need to build the home. Copies must be given to the building inspector for inspection and the financial institution for examination. Also, a builder should have three or four blueprints on the building site for the subcontractors to use.

Bow Casement Windows. A series of casement windows in one unit that curve outward. One or all may be opened depending on what you order. Inside screens cover the windows that open.

Breather Foil. An aluminum foil that is placed over the exterior sheathing or plywood before the siding is put on. It acts as insulation and repels heat. It is the first step in siding the home.

Brick Exterior. There are several styles of brick and they are sold in units or divisions of 1000. Generally, 1000 8-inch bricks will cover 143 sq ft. The cost of bricks compared to other types of exterior finishes is two to five times as much. The labor cost of building the exterior is also higher than installing other exteriors. Asphalt paper covers the exterior sheathing or plywood, then metal tabs are nailed into the exterior and bent to lay into the mortar between the bricks. The tabs are an additional support. If a brick exterior is going to be installed and you are going to have your windows capped in aluminum, have your bricklayers leave about ¼ in. around the windows. The aluminum installers need it to set their material around both sides of the molding.

have your bricklayers leave about ¼ in. around the windows. The aluminum installers need it to set their material around both sides of the molding.

Cable Connector. Used with Romex wiring, it fits into the knockout of an outlet box or switchbox and holds the cable in place.

Cable Strap. Used with Romex cable wiring, it is nailed to a stud to hold the cable in place.

Cafe Doors. They are decorative doors that open from the middle and swing both ways. Attached to the sides of the frame, they are available in a variety of lengths, styles, and woods. They can be stained or painted.

Canned-Goods Cabinet. This upright cabinet usually extends from the floor to the soffit. While narrow in width it may have one or two doors, one above the other, depending on the design.

Casement Windows. They open like a door via a rotogear, crank-type handle. The screen fits on the inside of the unit as the window swings open toward the outside.

Cast Iron Piping. By code, this is still used in underground installations. It is installed during rough-in and is covered by the basement floor. Other types of wastepiping can be connected to it once it reaches through the concrete basement floor.

Cedar Log-Cabin Siding. It has a curved exterior to match its name and is notched on top to lock into the siding above while grooved on the bottom to lock with the piece below.

Cedar Roofing Shingle. It is made from kiln-dried wood, and is very expensive. It should be used with specially designed homes incorporating it in the style.

Cedar Shakes. Among the most common sidings used today, it is also one of the least expensive. There are overlapping shakes in a variety of colors and styles. Some styles require painting every few years while newer styles have built-in color durability.

Cement and Brick. These two are the most expensive forms of siding depending on the area the home is built in. Not because of materials, but labor costs have increased the finishing prices.

Ceramic Tile (4 in. squares). It is generally used for bathroom walls but can be used in many areas. It can be applied to walls like plastic tile with one additional step. Grout, a form of plaster, must be placed between the tiles as a finishing step to cover the rough edges and act as a watertight sealing agent.

Ceramic Tile (1 in. squares). They can also be used on walls but are generally used on floors or vanity tops.

Ceramic Tile Trim and Moldings. These pieces include cap tile, down and up angle cap, base cove, and right and left cove angle.

Char-Glo-Grill (trade name). An inside barbecue unit that can be installed in the kitchen area. It is available in 19 in. and 29 in. width models. It comes with cover, adjustable heat control, brickets, and an optional rotisserie.

Chimes. The noisemaker of the doorbell available in a variety of tones and covers to match any decor.

Chopping Block. A 1½-in. thick wooden block that can be used to work on. The cost is comparable to Formica. It sits on top of any base cabinet. It is usually cut to order.

Circuit Breakers. They act as a safety device when shorted circuits are about to occur. Breakers don't have to be replaced. They have a switch that can be reset once the overload or short is corrected.

Clock Receptacle. A wallplate that hangs flush to the wall with a recessed contact point.

Closet Lock. A doorknob-type lock with a regular handle on the outside and a small twistknob on the inside.

Closet Rod. A circular hardwood pole used to hang clothes. It is usually made of birch and should be stained.

Clothesdrop. Made from galvanized square venting, it is shaped to fit between two studs. It may be a straight drop or it might require a small angled section.

Cold-Air Return Cover. A vent that sits off the floor in an interior wall. It acts as an air recycler by returning air inside the home back through the ductwork. The furnace sends the returned air out as new warm air.

Cold-Air Return Ducts. This ductwork receives the cooled air inside the home and recycles it back to the furnace.

Commitment Letter. A letter stating that a specific financial institution will issue a mortgage and final total payoff, but will not give any interim financing.

Concrete Blocks. The standard cement block dimensions are $15\frac{3}{8} \times 11\frac{5}{8} \times 8$ in. It is used in the construction of basements. In using this size block your basement walls will be over four inches higher than if you used modular blocks in an 11-course basement. The modular cement block dimensions are $15\frac{3}{8} \times 11\frac{5}{8} \times 7\frac{5}{8}$ in. This block is smaller than the standard block and is sometimes used in building basements with frost footings for attached garages. When facing a house with decorative brick this size of concrete block is more compatible.

Concrete Reinforcing Fabric (welded wire mesh). It is laid on the ground in the garage before the concrete floor is poured. Like the reinforcement rods in the footings, it adds strength to the concrete floor and prevents small cracks from growing larger.

Conduit Fittings. These accessories include locknuts, bushings, connectors, knockout seals, reducing washers, and reducing bushings.

Connector (90°). Used with conduit, it screws into a knockout hole of an outlet box or switchbox. The conduit clips into the open end of the connector and is set for the desired location.

Construction Fir. Lumber that is standardized and graded by government regulations. It ranges from eight feet to 24 ft in length in basic dimensions of 2×4 in., 2×8 in., 2×10 in., 2×12 in., and 4×6 in., etc. It is sold at all good lumberyards.

Copper Fittings. They include 45° and 90° elbows, tees, female and male adapters, pipe strap, and cap and union couplings.

Copper Tubing. This type of piping is used to draw water into the home and up to the faucets. It resists rust and corrosion and is very rigid in place and easy to work with.

Corner Base Cabinet. The cabinet is designed to fit into the corner and rest against the two walls. It has one door covering only part of the front. The remaining exposed portion will be butted to another cabinet at a 90° angle.

Corner Gap. Used to cover corners of the facia where two separate pieces join.

Covers. They come in a variety of shapes and sizes designed to cover outlet boxes. They have openings in them to fit over a combination of outlets and switches or individual units.

Crawl Space. An excavated area in the basement that is not completely dug out. The depth from floor to ceiling is usually between 3½ to 5 feet deep. The area is left unfinished and a hole is left in the concrete wall enclosing the area. This permits heatducts and other necessary pipes to be installed at the upper floor and later repaired if necessary.

Cultured-Marble Top With Bowl. A one-piece unit available with smooth or shell-shaped bowl, backsplash, and cutouts for faucet and stopper. Available in a variety of sizes including a two sink model. The two-sink models don't have a backsplash. A backsplash can be purchased separately.

Cultured-Marble Vanity Top. A level piece of cultured marble with a built-in backsplash and cutouts for sink and faucets. It fits on top of the appropriate base vanity. It's available in a variety of sizes, even large enough for two sinks.

Cultured Sod. The sod will vary depending on the area of the country you are building in. Sod originally grown on peat will not stay as green as sod grown on regular soil.

Damper. It consists of a circular head and straight tailpiece. The head fits inside the pipe while the tailpiece goes through an outside hole in the pipe then through the head. By twisting the tailpiece, the head is angled to allow more or less airflow through a particular run.

Decorative Brick. There is a wide variety of plastic and cement products in a vast assortment of colors and sizes. Plastic brick can be used on any wall where heat is not a factor. Cement bricks can be used anywhere because they are heat resistant.

Diagonal Base Cabinet. It's designed with one entrance door to fit into a corner base area. Regular base cabinets can be attached on either side. There are a variety of designs complete with lazy-susan shelves.

Diamond Grid. A removable wood decoration for a casement window. Available in a diamond or square shape, it clips to the inside of the window frame.

Dimmer Switch. It has a tap knob for ON and OFF and a rheostat that can be turned to adjust the brightness of the light.

Double Channel Runner. Used in areas where corners are joined and in some cases near the corners of the soffit.

Double Hung Window. A single unit with two windows that go up and down. It is the standard window used in the construction of most homes today. The heighth is longer than the width.

Double Joist Hangers. Same as single joist hangers only holding two studs side by side.

Downspout. It comes in ten-foot lengths and carries the water from the gutter down the side of a building.

Downspout Band. A three-sided brace or band nailed to a wall around the downspout to give it extra support and stability.

Drainfield. It consists of clay tile or concrete pipe usually four inches in diameter. The drainfield will be in varying lengths depending on soil absorption rates as determined by the percolation test. The drainfield is buried underground about two feet below the surface. Rocks are placed above and below the pipe. It is connected to the tank through a small concrete distribution box.

Drawer Base Cabinet. It comes with pullout drawers instead of standard swing doors.

Drive Strap. Similar to a nailup strap, only shaped to fit a rigid conduit.

Driveway Coating Sealer. A liquified-asphalt base that is brushed on the surface of the blacktop. It acts as a sealing agent and covers oil and grease spots.

Driveway Patch. Used to repair large cracks and potholes when blacktop is dry. It is as waterproof as the original blacktop.

Drop Siding. This horizontal siding has a flat surface and slants inward at the top to tuck under the siding above.

Duct Elbows (left-hand reverse; 90°; right-hand reverse). All of these are endpieces of square duct venting. Their use depends on the position of the opening to be fit between the joists in the basement ceiling and the first floor.

Duplex Outlet. A unit that has two independent outlets, one above the other, able to accept male plugs.

Dutch Doors. A particular style of door that is cut in half horizontally to allow half of the door to open.

Elbows. They are available in 60°, 75°, or 90° units to angle the downspout away from the three-inch drop on top. Also, they are attached to the bottom of the downspout to angle water away from the foundation.

Electric Baseboard Heat. Similar in design to a hot-water heating system, it has a heating element running through the center of the baseboard unit, instead of copper tubing with hot water flowing through it.

Electronic Filter. It filters the air from the furnace and frees the air of pollen, dirt, dust, smoke, and odors all year around. It should be cleaned once a month by soaking submerged in soapy water for 15–20 min then rinsed off with a hose.

End Caps. They cover the end of the gutter run.

End Panel. A single panel that covers one side of an exposed kitchen sink. There is a panel made to fit on either side.

Entrance Frame. It's available in single or double door units in combination with none, one, or two sidelights.

Extension. An extra piece of downspout connected to the main downspout via an elbow. It keeps the rain water from escaping too close to the foundation of the building. It need be only four to six feet in length.

Exterior-Plywood Siding. Available in texture, rough, grooved or rough cedar, and board and batten pattern facings.

Facia. The front of the soffit usually covered by the gutters. It's the board that covers the rafters and holds one end of the soffit in place. Facia board is made of wood.

Facia Cap. The aluminum or steel covering over the facia board.

Facia Runner. A molding nailed against the wall to hold the other side of the soffit in place.

Faucets. These have to be purchased separately from the sink. You can select from single or double sets and an optional spray unit.

Fiberglass Door. Four heavy-gage fiberglass panels mounted into an aluminum frame. Both panel and frame can withstand any weather. It never needs painting and natural light can be seen through the door. Individual panels are replaceable and are of very light weight. They are available in a variety of colors in standard and special sizes.

Fiberglass Kraftface. Available in widths from 15 to 23 in. and thicknesses of 2¾ to 6½ in. This insulation is placed between the studs in the exterior walls and between the joists in the ceiling. It is stapled to the studs and joists to stay in place. The fiberglass paper covers the outside exposed area.

Fiber Insulation (backerboard). It fits into a pocket area behind the siding to give support to the aluminum or steel and acts as an additional insulation factor.

Fittings. These are the accessories needed to bend pipe or make extensions. The names and shapes are standard with the type of piping material used.

Fixture. A light holder on a wall or ceiling installed during the finishing period.

Flashing. A small metal square used around and up the side of the chimney and under shingles. It seals joining areas to prevent water from coming into the attic. It can also be used on one edge of the roof where two floors join together.

Flexible Steel Conduit. A hollow tube that can be bent and turned to almost any position. It runs through the holes drilled in the studs and joists from outlet box to switchbox. The electrical wire is then run through the tubing from connecting switches to the outlets. Greenfield is the name of the wiring system using the conduit and separate wiring.

Flexit Doorbumper. A coil-type doorbumper that moves or gives when a door bumps against it.

Floor Drain. A floor outlet connected to the sewer pipe that will accept water. The number of drains used must be determined when the home is in the blueprint stage. The concrete men must slant a portion of the floor around the drain to make sure any water on the floor will fall toward the drain.

Floor Register (with diffuser). A flat register and cover for a warm-air duct. The diffuser closes tight or permits as much warm air in the room as you want it to. Also, it should be placed on the floor in front of the permanent glass area of a sliding patio door.

Folding Doors (with frames). They are usually made of plastic or vinyl and are accordion pleated. They hang from an overhead track and are available in one- or two-door sets.

Footings. This is the first step in the construction of the basement. Concrete is poured between wooden forms. When dry it acts as a base to support the concrete block walls, steel columns, and masonry chimney.

Forced-Air Furnace. This power plant in the home functions the same way regardless of which energy source is used. It uses a ductwork service to distribute the warmed or cooled air throughout the house. You must choose the power source according to what you prefer and what is available.

Formica (trade name). Available in a textured or flat surface. The thickness depends on the style desired. A thinner style can be molded and preformed while the standard thickness cannot be bent without breaking.

Formica Countertop. It has a standard Formica top. The edges are squared and it has no backsplash. A separate backsplash can be added by gluing a four-inch wide Formica strip to the wall behind the countertop. The top, bottom, and sides are seamed with a metal overlapping strip.

Foundation Coating. A liquified asphalt used to waterproof the outside of the foundation blocks that will be covered by backfill.

Fuses. The older form of safety device. The fuse "blows" and must be replaced if a particular circuit is overloaded or shorted.

Fuse Adder Box. It adds additional electrical circuits to the home and is attached to the incoming electrical cable, usually next to the original service box.

Fuse Box. It contains the fuse-type circuits for the electrical service.

Galvanized Fittings. These include 45° and 90° elbows, tees, union caps, reducing couplings, bushings, and cored plugs.

Galvanized Pipe. A steel or iron pipe coated with zinc to prevent rust. It is not as flexible as copper tubing but less expensive. The larger diameter that is available can be compared to copper tubing. Check your local code to see which can be used.

Galvanized Pipe Connector. A small piece of piping used to splice circular ductwork together.

Galvanized Roof Edge. It covers the edge of the roof not covered by shingles and prevents the wood from weathering.

Galvanized Roof Valley. A curved strip of metal used to cover adjoining areas where the inside corner is formed on the roof. It prevents water from leaking in.

Galvanized Tee. Attached to the furnace, it acts as a splice in a duct line by branching in two directions.

Galbanized Venting Pipe. A piping used to create sideruns from the main duct. It is circular in shape.

Garbage Disposal. It connects to the drain at the bottom of the sink and grinds garbage for easier disposal. It operates on a separate switch with cold water running through the disposal while it is working. It's available in many sizes. Usually, the disposal is installed on one side of a two compartment sink while the main drain is in the other compartment. The adjacent sink is connected by a separate pipe to the main drain system.

Glider Windows. They are built similar to a double hung window only they open by sliding sideways. The width is longer then the heighth.

Greenfield. The type of electrical wiring system combining conduit tubing with separate interior wires from switch to outlet.

Grille. A removable wood decoration for double-hung window units. Available in a variety of styles, it clips to the inside of the window frame.

Grooved Facia Boards. Available in 1 × 6 in. or 1 × 8 in. boards that cover the front of the overhang or eve. Grooved on the inside lower portion the board, it holds the plywood soffit in place while the other end is permanently nailed perpendicular to the outside wall of the home.

Grounding Clips or Screws. The clips or screws used inside switchboxes or outlet boxes to ground electrical wiring.

Gutter. Available in 10-foot or 20-foot lengths in prefinished aluminum or galvanized iron.

Gutter End Piece (with three-inch drop). A short accessory containing a hole with a three-inch drop in it. The downspout connects around the drop to permit water from the gutter to escape down into the downspout.

Hangers. They are nailed to the roof at one end while the other end supports the gutter.

Hardboard Siding. This horizontal siding has a smooth surface. It is nailed to the exterior in an overlapping fashion. A primer coat usually covers both sides.

Hardwood Plywood Flooring. This flooring is ¾ in. thick, comes in 4 × 8 ft sheets, and is used as the subflooring nailed directly over the joists. It is installed during the rough-in period.

Header. A horizontal piece of wood forming the top side of the doorway or frame.

Hollow-Core Door. It is usually constructed of plywood on either side of a frame forming a hollow center. It's not practical for use as an outside door in cold climates as there is little insulative quality to its structure. Windows can be cut into it and it's available in a plain or fancy exterior. It could be used for all or most inside doors.

Home Specification Booklet. A written breakdown of the various building materials that will be used in the construction of the home. The financial institution and the building inspector both need a copy.

Hot Water Heat. A baseboard system recycling hot water through copper tubing. It requires a small water tank attached to a furnace. No ductwork is needed. The only drawback in this heating unit is that central air-conditioning cannot be hooked up with it unless a separate ductwork system is installed during the rough-in period.

Humidifier. This unit adds moisture to dry winter air. During the cold months the humidity control knob should be adjusted according to the outside temperature for maximum effectiveness. Heated air is drawn from the furnace into the unit and forced through water-saturated pads. The moist air is sent back through the ductwork to be distributed through the home.

Inside Corner. A 90° angled strip used to cover all inside corners of the outside walls.

Inside Dump Pump. Placed in the basement of a home with a high water table around the house, a hole is cut into the concrete floor and partially filled with crushed rock surrounding the pump. Water is permitted to come through and, when it reaches a certain level, the pump is activated. The water is forced up through a hose to the nearest drain. It continues to work until the water drops to a safe level.

Inside Gutter Miter. A 90° corner piece used to turn the gutter around the inside corners.

Insulated Thermal Door. A heavy unit that has a polyfoam type of insulating interior with a hardboard or steel exterior. Windows are available in some models together with a wide variety of designs in four to six sections with single or double doors.

Insulation Board Sheathing. It's available in 4 × 8 ft and 4 × 9 ft sheets of ½ in. and 25/32 in. thicknesses. Applied only to exterior wall studs, it acts as a base for the facing material, provides an extra layer of insulation, and does the job it's designed to do. It does not have the strength of plywood sheathing.

Insulators. They are connected to the outside walls or to the service entrance pole. The outside electrical wires are looped through the insulators before they enter the service head in the pole.

Interest. The amount of money the financial institution charges the borrower for his loan.

Interim Financing. A short-term loan issued by a financial institution giving an individual money to pay the subcontractors at the completion of the rough-in stage and other designated times. One is only charged interest for the amount of money that he withdraws.

Inverted Register Boot; (90° angle body; 90° end boot). All three are endpieces of circular venting. They are used, depending on the position of the opening, to fit between the joists in the basement ceiling and the first floor opening.

Jamb. A vertical piece of wood forming the sides of the doorway or window frame.

J-Channel. Shaped like a *J*, it is used around windows and doors.

Joint Compound. When mixed with water it forms a cement-like thickness. It should sit about 24 hours before additional coats are applied to the plasterboard. It is used to cover all nail indentations, seams, and inside and outside corners of two adjoining sheets of plasterboard.

Joint Tape. Used to cover seams of adjoining sheets of plasterboard. A thin coat of joint compound is applied before the tape. Two additional thin coats of compound are used as finishing coats over the tape.

Joists. Similar to the studs, they come in longer and wider boards. They are used in the rough-in construction for the first floor, second floor, ceiling, rafters, collar ties, and valley and hip rafters.

Knockouts. These are precut covered holes that can be punched out. They are located on the sides and back of the outlet boxes or switchboxes. They are used to attached conduit to the box or thread Romex wire to the clamp.

Laminated Plastic Top. Similar to a vanity top and complete with backsplash, it usually has a particleboard interior and is covered with laminated plastic or Formica.

Lattice. Oak or pine trim usually ¾ to 1½ in. wide. It comes in lengths up to 20 ft long and is used to cover the seam caused by two sheets of paneling nailed together.

Laundry Tubs. Today's tubs are made of a variety of materials from plastic to steel. Determine the type and length of wear you are planning to give the tub, then make the proper selection. They are available in single and double units. Some units can be hung on the wall, but most units sit on a stand with adjustable leveling screws on the legs.

Lavatories. They are usually made of china, very inexpensive, and used if a vanity is not necessary or space is not in quanity. Some units come with faucets.

Lavatory Hardware. These small finishing accessories include towel bar, toothbrush holder, tumbler holder, soap dish (recessed or wall mount), robe hook, paper holder (recessed or wall mount), and towel ring.

Left Hand Door. It's a door that opens from the left side when you face it.

Liabilities. These are obligations one owes that require a financial settlement.

Lien Waiver. A form of receipt that is completed and signed by a subcontractor or distributor when he is paid for his labor or material.

Lockset. The exterior doorlock and pull combination.

Long Reducer (big end crimped). The crimped end fits into a circular venting line while the small end is free to accept a circular vent of a smaller size.

Long Reducer (small end crimped). The crimped end fits in the circular venting line while the large end is free to accept a circular vent of a larger size.

Louver Panel. Available in a variety of shapes and styles, the panels you select will be determined by the style of home you build. Their main functions are to permit air into the attic, keep moisture buildup out in the winter, and let the warm air out in the summer. They may be placed on the roof, gables, sidewalls, and in the soffit.

Masonry Chimney. Beginning in the basement, it goes up through the home and roof. It is constructed of special concrete blocks and piping, and used to vent the gas furnace, water heater, and incinerator.

Medicine Cabinets. The styles and sizes will vary, but there are two basic types—surface mount and wall mount. The surface mount type is supported by wall screws that are screwed right into the wall. The wall mount type requires part of the wall to be cut out between the studs where the cabinet is inserted, leaving just a small lip outside the wall. It is screwed into the studs for support.

Metal Door. It has a polyurethane core bonded to a galvanized steel shell. This eliminates the need for a storm door. This door is resistant to warping, bending, shrinking, and swelling. It is excellent for a front door or connecting door between the house and its attached garage. Metal doors are available in a wide variety of styles, colors, and with windows.

Metal Tile. Available in a flat or brushed-finished face, it is usually found in the kitchen around the stove area. Grease wipes off and can't penetrate the surface.

Meter. A gage that measures the amount of electricity used in your home. It is connected to the service entrance conduit or cable.

Mineral-Wool Furring Blanket. This is thin insulation, in 8 ft lengths of $1\frac{1}{2} \times 15$ in. that fits between the studs and joists.

Mortgage. The final contract and payment schedule a borrower has with a financial institution covering the balance of money he still owes.

Nailup Strap. *J*-shaped, the long edge is jagged and nailed into a stud or joist. The hook of the *J* holds the conduit in place. It's used with flexible conduit.

Oak Base. The hardwood trim that should be used on the exposed wall of the house as it can withstand most scratches. This base holds a better stain. It is more expensive than pine base.

Oak Casing. The outside finished trim that surrounds the door opening. One should use a hardwood such as oak, as opposed to a softwood like pine. It takes the best strain and resists scratches.

Oak Flooring. This is the most expensive finished flooring. It must be put on in individual pieces and sanded to a smooth shiny finish. It should be used in rooms where the floor will be partially or fully exposed.

Octagon Box. It has four long sides and four curved sides and is installed during the rough-in period. There are several different styles with a variety of inside clamps and outside brackets, depending on the style of outlet used.

Octagon Window. An eight-sided small window that can be set almost anywhere. It's a one-window unit sectioned off by two vertical and two horizontal bars.

Outlet Box. A four- or eight-sided box that acts as a holder or interior brace for an outlet. It is attached to a stud during rough-in at a specifically designated height.

Outside Corner. A 90° curved piece used to cover all outside corners of the home.

Outside Frame and Sill. The sill is the same as the base and it lays next to the threshold on the outside of the house. The frame comes in a standard 4¾ in. or oversized 5⅛ in. width. The oversized frame is used when paneling is placed over a plasterboard wall. The frame consists of two 6 ft 8 in. inside casings and one header made of pine, mahogany, or oak. The header is the top piece of the frame while the casings are the sides.

Outside Gutter Miter. A 90° corner piece used to turn the gutter around the outside wall corners.

Oven Cabinet. Designed to accept a built-in oven unit, at a desired height, and has storage areas above and below the oven unit.

Paneling. A wide variety of decorative, rough, and smooth-surfaced sheets of wall coverings are available in a variety of thicknesses and lengths. The most popular size is 4 × 8 ft and is ¼ in. thick. They are available in wood and wood byproducts. The prefinished surface can make or break the design of the room.

Passage Lock. A door latch for an inside door. It cannot be locked.

Particleboard. Available in ⅜ to ¾ in. thicknesses in 4 × 8 ft sheets, it's made of wood flakes and molded into a smooth, dense board. It is used as the top layer of the finished floor where carpeting will cover it. This material should not be used under linoleum as its surface is too smooth to grip the linoleum glue.

Perforated Panel. Used in the soffit for ventilation every few feet.

Personal Statement. A form the financial institutions request a borrower to fill out in order to inform them of his liabilities and assets.

Picture Window. Usually, it is the center section of a three-window unit. The large window is made of one-inch insulated glass and the two side units are double hung or casement windows.

Pine Base. It can be used in closet areas and other places that will not be in view. The stain usually smears and looks unclean.

Plaster Corner Piece (8 ft). A metal perforated strip, with two sides fixed at a 90° angle, used to cover the outside plasterboard corners. It is nailed in place and joint compound is applied over it.

Plastic Fittings. They include coupling, sanitary tee, vent tee, closet bend, 90° elbow, 45° elbow, 45° Y reducing flange, male adaptor Y, bend combo hub, reducer bushing, tailpiece with nut, and spigot.

Plastic Pipe. In many areas plastic pipe has passed the inspection code and can be used anywhere above ground to accept waste and send it out of the home. It is easy to install, light weight, and flexible. It can be cut with a knife or saw. It resists corrosion and accepts any fluid that can be washed down any other type of drainpipe. The smooth inner walls allow no internal buildup of waste material or the growth of bacteria. The only bad feature is the hollow sound you hear when the water runs through it. Insulation should be packed around the pipe and hidden in the walls during the rough-in stage to prevent excess noise.

Plastic Tile. It can be used in any area of the home where heat is not a factor. It is used mostly on bathroom walls, is very inexpensive, and is available in four-inch squares.

Plate Rail. A decorative rail that fits on top of a wall cabinet that is not hung from a soffit. The top of the cabinet will be hung one foot below the ceiling. This space can be used as a display area.

Plywood Flooring. Available in ¼ to ½ in. thicknesses and 4 × 8 ft sheets, it is used as the finished floor that will be covered by linoleum. The plywood surface is very porous and holds the linoleum glue for an indefinite period.

Plywood Sheathing. Available in 4 × 8 ft sheets in thicknesses of ⅝ to ¾ in. It is the first layer of material to be applied to the studs during the rough-in framing of the house. It is applied to form the exterior walls and the roof. Plywood lends more support to the structure than insulation board but it doesn't hold the heat in the home as well.

Pocket Door. A sliding door unit. The frame is built during the rough-in period of the house. The door slides into the wall on an overhead track that is prefastened to the header or the top of the finished door frame.

Polyethylene Sheeting. Available in a variety of depths from 3 to 20 ft and in a thickness from 2 to 6 mils. It is excellent as a vapor barrier and an all-purpose covering. Usually, it is stapled over studs and joists after the insulation is put in place in the home.

Polystyrene Insulation. Available in ¾ to 2 in. thicknesses, it is 8 ft long and 14⅜ in. wide. Usually used in the basement, it is light weight and easily glued to the cement block wall, between the 1 × 2 in. stripping, 16 in. on center. It is waterproofed and acts as its own vapor barrier. Paneling can be placed directly over it.

Pouring Wool. Loose insulation that is poured between the joists in the attic.

Preformed S-Curve Countertop. A thin layer of preformed Formica with a four-inch curved backsplash against the wall and curved edges in front. The Formica is glued to a particleboard base.

Prehung Door. The door is permanently attached, via hinges, to the jamb. It is usually a hollow-core door made of oak, pine, or mahogany.

Privacy Lock. A door lock for the outside door.

Ranch Windows. These are small rectangular windows used mostly in the garage area.

Recessed Ceiling Fixtures. These units, usually prewired, contain their own outlet box. The box is nailed to a wall stud or ceiling joist so that the fixture cover, when attached, lies flush with the ceiling or wall.

Red Rosin Paper. This paper is laid between the subfloor and the finished floor. It prevents floor noise and absorbs house moisture.

Redwood Siding. This versatile siding can be positioned on the exterior in either a vertical or horizontal fasion, and can be left in a natural state or stained.

Register Head. A rectangular opening piece installed between studs during rough-in. It is used as the endpiece of the cold-air return system. The cold-air return cover screws into the head.

Reinforcing Rods (rerods). These steel rods are placed in the wooden forms of the footings before the concrete is poured. They add strength to the dried concrete and prevent the eventual cracking of the footings from getting too large and weakening the base.

Romex. The type of wiring system that has the wire enclosed in a thick polyethylene coat. It will not corrode or rust and is flexible. It works on the same principle as flexible steel conduit. The wire is routed through holes drilled in the wall studs and joists.

Roof Coating. The coating acts as a waterproofing cement. It is usually not applied to roofs of new homes.

Roof Flashing Collar. This unit seals the service mast on the outside of the house as it goes to the roof.

Rough Cedar. A rough-faced cedarboard siding can be placed on the exterior in a vertical position. Batten is placed over the area where boards join together.

Rough-in Wastepipe. An open pipe extending out of the cement floor in the basement. It is installed during the rough-in stage for a future toilet in the basement. It is connected to the sewer pipe below the floor. Until it will be used, you should stuff rags about four inches into the pipe, then cap it with cement until needed. This step will prevent sewer gas from backing up into the basement.

Rough Sawn Cedar. It has a natural color with a rough surface. It is used on the exterior as a siding material and inside the home to make decorative ceiling beams. It can be stained to a desired color. It's available in a variety of lengths and widths.

Safety Switch. It's connected to the main power line and can shut off all the power to the home.

Sash Lift. A handle used to lift and lower the window.

Sash lock. A window lock.

Seepage Pit (cesspool). A tank buried in the ground outside the house that accepts the water drainage from the septic tank. It may or may not be needed depending on local code regulations.

Self-Storing Storm Windows and Screens. They are available in a wood or aluminum frame. The top outside area is glass and rests on a screen. A sliding storm window is on the inside to cover the screen area.

Septic Tank. A solid concrete (or coated steel) tank buried in the ground outside the house. This tank is the first unit hooked up to the house, and accepts and stores most of the solid waste. The sizes vary and the local code regulations will inform you of the proper size for your home.

Service Box. The metal box located inside the house where all inside electrical hookups meet. It contains circuit breakers or fuses depending on the system installed. The outside power source is connected to the service box.

Service Entrance Cable. The Romex-type cable coming into the house, through the meter box and wall, to the service box.

Service Entrance Conduit. A straight hollow tube that goes up the outside wall of the home from the service box through the meter box. It forms a protective shell that contains the wires bringing the electrical power to the home from an outside power source.

Service Mast. A pipe and its accessories extending from the roof, going down the outside wall, that accepts the electric cable from the outside power source. It runs through the meter and wall and into the service box.

Service Mast Components (for cable). These components are the service entrance head, cable strap, service entrance cable, and weatherproof connector sillplate.

Service Mast Components (for conduit). These components are the service head, conduit, conduit strap, and weatherproof adaptor entrance.

Sheetrock (tradename). A type of plasterboard available in a variety of thicknesses and sizes. Usually, one uses a board that is ¼ in. thick and 4 × 12 ft. It is raw wallboard nailed directly to all interior sides of the outside stud walls, to joists in the ceiling, and to both sides of the room divider or internal stud walls, after the rough-in of all subcontractors is completed. This product is used in place of plaster walls.

Shelving. It can be made of any spare lumber, but it should be a hardwood. Also, what is to be stored and the length of the shelf are factors to consider when selecting the type of wood.

Shower and Bath Fittings. These are the faucets for the tubs. They are available in a variety of shapes and sizes and purchased separately from the tub.

Shower Stall. Usually this type of unit will be installed in a bathroom that is added after the home is completed. It is available with plastic sides or enamel covered steel. Individual units can be purchased with a door or curtain and with or without fixtures. You must determine if you want faucets on the left or right side of the walls. The drain in the floor of the stall must be connected to a drainpipe in the basement. The drainpipe should be installed during the plumbing rough-in of the house.

Shutters. Available in sizes to match most windows and doorways, they come in a variety of designs in aluminum, wood, or vinyl material.

Side Lights. They are see-through or opaque windows permitting light to enter on one or both sides of the door, depending on the number of windows. They are usually placed at the main entrance of the home.

Side Wall Register. A vent, usually placed by an entrance way, off the floor. It's complete with a diffuser to shut off the vent or angle it to the desired direction.

Sill Cock or Hose Bib. This unit is on the outside wall of the home. The garden hose is connected to it for outside water usage. An inside turnoff can also be installed.

Sill Sealer. Available in 50 ft rolls with a 1×6 in. thickness. It is laid over the upper course of basement blocks before the first piece of lumber is put into position over it to start capping the house.

Single Joist Hanger (metal). A three-sided U-shaped piece of metal that is nailed 16 in. on center to the center support beam of the floors and ceilings. It is used to support the inside end of the support joists and the beams that are placed across the walls.

Sink and Range Cabinets. Available in one or two door units and designed to accept a sink or cooktop. It has no storage shelves, as pipe connections take up some of the inside area.

Sink Front. A single panel that covers the front area of an exposed sink.

Sliding Patio Doors. They are available in aluminum or wood frame with insulated glass. Double window units have one gliding door, one fixed door, and a sliding screen. Three and four window units are also available. Outside key locks are available on all units. A wood frame is advisable over an aluminum frame in cold climates.

Slip Head Window. A single window that, when raised, goes through an opening between the top of the casing and into the wall.

Slipjoint Connection. It joins two gutter pieces together.

Soffit. The underside of the roof overhang usually constructed of the same material that is used in siding the house. But many people blend wood with a variety of sidings. Regardless of material used, breather spaces should be used every few feet to allow for additional airflow in the attic. Some aluminum and steel soffits are self aerating, depending on the manufacturer.

Solderless Connector. A screw-on plastic cap that twists over the connection of the wire from the fixture and the wire from the outlet box when they are twisted together.

Solid-Core Door. This door is made of solid wood. It may be plain or a design may be cut into it. It usually is made of oak, birch, or mahogany.

Space Heaters. These independent units vary from a small electrical plug-in model (to heat a few square feet) to a cut-down version of the main furnace (to independently heat the basement). All gas-operated units should be vented out through the stack or an outside wall. Electrical units should be placed where the plug-in cord may be concealed.

Spike. A large nail-like piece of metal that is cemented into the course of the foundation walls every few feet. It is used to anchor the first board put down on the top course to cap the house.

Square. Shingles are sold by the square. A square is the number of shingles required to cover a flat 10 × 10 ft area of a roof. The numbers 235 lb, 245 lb, etc., refer to the weight of the shingles that make up one square.

Square Box. This outlet box has four sides. Installed during rough-in, it comes with a variety of inside clamps and outside brackets depending on the style of outlet used.

Square Vent Duct. Used as the main passage way from the furnace to the various runs through the home.

Stair Riser. An individual piece of wood cut to size and nailed in place from one tread to the other that acts as a back stop.

Stair Tread. An individual piece of wood cut to size and nailed in place to form the part of the stair unit that one steps on.

Starter Roll. It goes on the edge of the roof first, just above the facia. It is used with standard 3-tab asphalt shingles. These shingles have two narrow opening slits that would expose the asphalt paper on the first row only. The starter roll fits under the shingles covering the asphalt.

Starter Strip. The first siding panel put all around the home at the base of the walls above the basement.

Steel Beam. The beam lays inside the top course of concrete blocks across parallel walls. It is used to support inside weight-bearing walls.

Steel Column. A vertical, adjustable, steel beam support that is based on a subfloor concrete footing. The concrete basement floor is poured over the bottom four inches of the column to add additional support.

Steel Nails. They are used in conjunction with steel siding.

Steel Siding. Available in a variety of thicknesses and colors in a smooth or embossed finish, it can be used with almost any other siding. It must be cut with a hand tool to prevent sparks from burning and sticking to the finish. It is available in vertical and horizontal lengths.

Stepping. The two sawframes that form the stairs. The treads and risers lay across them and are nailed to them.

Stone Exterior. Depending on the styles of stones selected, it is expensive to buy and build.

Storage Fronts. The door and molding unit used to cover a small storage closet.

Storm Door. It is placed outside the entrance way of the home. It's constructed of metal or wood in a variety of designs including self-storing screens and storm window combinations.

Straight Boot. This endpiece fits on the end of a circular venting line that doesn't require any twists or turns.

Strainer. A small mesh screen placed over the hole of the three-inch drop to prevent leaves from clogging the downspout.

Strap. A U-shaped holder for rigid conduit. It is nailed to a stud or joist around the conduit to keep it in place.

Stucco. A plaster-type exterior that requires asphalt paper over the exterior sheathing or plywood, and wire mesh nailed over the asphalt paper. The material cost is low but labor cost will bring the price back up. It can be blended with almost any other type of exterior product.

Studs. The common name given to basic 2 × 4 in. kiln dried wood used in framing the walls of the house. It comes in various lengths, the standards ranging from 8 to 12 ft.

Sump Pump. A motorized pump used to draw water out of a drainage sump pit to an above drain outlet close to ground level, since water will not flow in an upward dirction.

Switchbox. It is rectangular in shape and installed during rough-in with a variety of inside clamps and outside brackets to fit a specific type of wall switch.

Switch/Outlet Combination. A single unit that has one female outlet on the bottom and a switch on top.

Switch/Outlet Plates. Decorative covers that fit over a unit leaving exposed the switch or the outlet. They fit flush to the wall.

Sworn Construction Statement. A form used to list the chosen subcontractors, their charges, and the schedule of payments.

Thermostat. A gage that regulates and maintains the desired temperature of the home. It is directly connected to the furnace and signals the furnace to go on or off automatically for desired heating or cooling.

Threshold. A preformed wood base nailed to the floor that the door rests on when it is closed. It is made of wood with either a vinyl or metal door-rest over it.

Tile Accessories. These accessories include the paper holder, soap holder, towel bar, and toothbrush holder.

Tileboard. A precut panel with a permanent baked-on finish that can be used in the bathtub area and kitchen. Additional accessories include tub edging, inside corner molding, and cap strip molding.

Toilet. It is available in a wide variety of sizes, styles, colors, and prices depending on what is needed to blend with the other bathroom fixtures.

Toilet Seat and Cover. This two-piece unit covers the open area of the toilet. The basic popular types of material that the units are made of are enameled wood, plastic, or marbelized plastic. There are many styles and colors from which to choose.

Ton of Cooling. A measurement of air-conditioning needed to cool a specific area. A ton is equal to 12,000 BTU (British Thermal Units) which will usually cool 600 sq ft of floor space very adequately. When cooling a multilevel home this formula may vary slightly.

Trap. The pipe that extends downward from the sink through the floor to the sewer. It permits waste to flow down from the sink.

Trees and Shrubs. These items, like sod, will vary depending on the area of the country and the type of soil that will support them.

Trim or Molding. A strip of wood of various designs used to cover the area where the wall cabinet joins the soffit. It is a finishing trim.

Tub Enclosure. A wide variety is available and can fit any style of tub constructed today, including recessed models. They work on a sliding door concept and can be purchased with frosted glass, frosted plastic, wire-tempered glass, or clear glass. The safest models to use come with tempered glass or hi-impact plastic. The frame is made of aluminum and has rubber bumpers.

Tv Outlet. A special outlet designed to accept the wiring from your outside TV antenna. The adaptor cord fits into the outlet and connects to the TV set.

Under Sill Trim. It fits the sills of the window and must be cut on a machine, on the job, from a flat piece of aluminum. The shape will depend on the style of sill used.

Utility Cabinet. Similar in looks to a canned-goods cabinet. It may or may not have any shelves depending on the storage purpose.

Valance. A decorative piece of trim, usually over a window area, joining separated cabinets to give a continuous look.

Vanity Drawer Base-Cabinet. A small base cabinet containing pullout drawers instead of doors.

Vanity Front. A false-front panel used to cover the front of a vanity top.

Vanity Hamper (base type). It has a small pullout drawer with a larger base hamper for storage of soiled towels and clothes.

Vanity Panel. A false-front panel used to cover one or both sides of a vanity top.

Vanity Sink. Available in enameled steel or porcelain, it may be purchased to fit into a vanity top or stand by itself without a base cabinet.

Vegetable Base Cabinet. A standard cabinet in looks, but the drawers are designed to pull out and downward to place and hold vegetables.

Vent-A-Hood (trade name). A necessary unit to pull smoke out of a kitchen when a *Char-Glo-Grill* is installed. It is available in a variety of styles, colors, and sizes to fit every kitchen decor. It must be vented out through the ceiling. The hood should set no more than 27 in. from the grill or cooktop for maximum drawing power.

Vent Fans. These units should be hooked up in the bathrooms. They are required by code if no outside ventilation is available for the bathrooms.   If possible, every bathroom and kitchen should have one. There is a wide variety available and some have additional lights, sunlamps, and heating units with them for multipurposes.

Vent Saddle (45° or 90°). The saddle-shaped base fits over circular venting. Screwed into position, a hole can be cut into the main pipe and the air can be vented into the desired direction.

Vermiculite. A handpoured insulation used between the exterior wall studs.

View Vent Window. This unit is a combination of a larger top nonopening window with a bottom awning unit. A storm window is available for the upperglass area or an optional half-inch insulated glass is used. An inside screen covers the awning unit.

Vinyl Channel. The track inside the window unit that allows them to slide up or down. There is less heat lost using a vinyl channel than an aluminum channel.

Wall Cabinet (corner). A single or double door unit designed to fit into corners. Regular wall cabinets can be attached to either side.

Wall Cabinet (soffit). This cabinet fits between the ceiling and the regular wall cabinet where a soffit would usually exist. It has sliding doors and is great for extra storage space.

Wall Cabinet (standard). These cabinets hang on the wall and under a soffit, if any. They are available in one or two door models in a variety of standard widths and depths.

Wall Corner Unit (diagonal). The same as a standard wall cabinet that lies flush against the corner and against the sides of the two walls. And additional wall cabinet can be attached to only one side.

Warm-Air Duct. This ductwork comes directly from the furnace and circulates the new warm air through the home.

Water Heater. These units are measured in gallon-size amounts. For today's living a 50-gallon hot-water tank should be the minimum size, when you consider all of the modern units (such as a dishwasher, clotheswasher, etc.) that consume hot water. The actual cost of a large tank is not that much greater than the minimum 30-gallon size. The cost of installation is the same regardless of the size. Water heaters can be purchased in gas powered or electrical units. For a longer lasting heater choose a unit with a glass-lined tank.

Water Softener. A system that changes hard water into soft water. It connects into the main incoming waterline. You should check the type of water available in the area you are building to determine if this device is needed.

Weatherstripping. There is a wide variety for every purpose, but it is usually used around windows and doors to prevent excess air from entering the home.

Wood Sectional Door. Individual hardwood sections shouldn't crack or warp if properly weatherproofed by painting. The single and double door units are available in a variety of styles in four or more horizontal sections. They tend to absorb humidity and are heavier than normal when wet. Windows are available with these doors.

X-Bridging. Wood strips nailed from the top of one floor joist to the bottom side of the adjacent joist. The bridging crosses between the joists in the shape of an X. The bridging adds to the support and strength of the floor joists. It must be used under every floor.

14

Pictorial Review
of Construction

This is the final chapter of a book about building a house that became just as involved as the actual building of the house itself. In this chapter you will see a pictorial review of the construction phase starting from the shoring up of the basement walls to the completed house.

Previous chapters have covered just about everything you should know about subcontracting the building of your own dream house or where to find additional information. This chapter shows you how the actual construction should progress, step-by-step, phase-by-phase.

Of course, 36 photographs cannot show every single step of home construction, but every effort has been made to include a good representation of each construction area and every subcontractor's job.

Once the plans are made, blueprints are drawn, contractors are scheduled, and materials are ordered, the best part of your project will begin—watching your dream come true. This is ideally the most exciting part of the entire project.

Fig. 14-1. Completed basement walls are shored up in preparation for backfilling of dirt against the outside of the walls.

Fig. 14-2. To aid in waterproofing the basement, a polyethylene liner should cover the outside of the concrete block wall in addition to the waterproofing asphalt.

Fig. 14-3. After a wire mesh is laid on the ground in the garage area, a four-inch cement floor is poured. The mesh acts to prevent cracking from expansion and settling.

Fig. 14-4. Prebuilt roof trusses.

Fig. 14-5. Steel beams and columns in basement are positioned in place to support the main floors. Forced-air venting is positioned alongside the beams.

173

Fig. 14-6. Completed rough-in walls are built to scale on the job and positioned to be raised into place.

Fig. 14-7. Rough-in frame of main floor.

Fig. 14-8. Exterior sheathing covers the rough-in frame.

Fig. 14-9. Roof trusses are positioned and nailed in place.

Fig. 14-10. Exterior plywood sheathing is installed over the roof trusses.

Fig. 14-11. Completed rough-in roof.

Fig. 14-12. Partially finished roof shows exposed asphalt paper over plywood sheathing, and roofing shingles over the asphalt paper.

Fig. 14-13. Completed roof indicates exposed metal flashing where two sides meet at a specific angle. The metal acts as a water barrier to protect the home.

Fig. 14-14. Breather units are placed in the roof to permit hot air to escape from the attic in the summer and keep it dry in the winter.

Fig. 14-15. Breather vents are placed in the underside of the overhanging eaves to act as an additional aid in regulating attic heat and moisture.

Fig. 14-16. Before windows are installed in the rough-in walls, asphalt paper strips are placed on the sides. Asphalt paper will eventually cover the entire exterior-wall sheathing before the final siding material is installed.

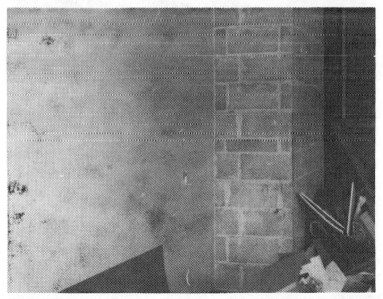

Fig. 14-17. A ⅝ inch thick plasterboard firewall is the common wall between the living area of the home and the attached garage.

Fig. 14-18. The rough-in interior frame for the ceiling and kitchen soffit.

Fig. 14-19. Interior rough-in walls with plastic plumbing inserted through predrilled holes in studs.

Fig. 14-20. Rough-in waste plumbing in the basement attached to the kitchen and basement wastepipe.

Fig. 14-21. Metal stack is placed in an interior rough-in wall. Stacks may also be made of special concrete blocks.

Fig. 14-22. Typical gas-fired forced-air furnace attached to ductwork.

Fig. 14-23. Heating and cooling forced-air ductwork installed within a rough-in interior wall.

Fig. 14-24. Rough-in forced-air return attached to an interior rough-in wall.

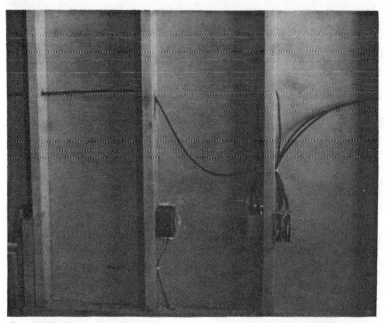

Fig. 14-25. Rough-in electrical wiring running through predrilled holes in the studs and connected at the outlet boxes attached to the studs.

Fig. 14-26. Exterior rough-in wall with insulation placed between the studs.

Fig. 14-27. The lower level of a split-entry home with insulation placed between the studs of the exterior rough-in above-ground wall.

184

Fig. 14-28. A polyethylene water barrier protects the insulation between the studs of an exterior wall and the plasterboard interior wall. The rough-in soffit hanging from the ceiling is ready to be covered with plasterboard.

Fig. 14-29. Plasterboard is being nailed to the studs of an interior rough-in wall.

Fig. 14-30. Rough-in wall partially covered by plasterboard.

Fig. 14-31. Raw plasterboard completely covers the rough-in interior wall.

Fig. 14-32. Finished plasterboard with seams and nail holes covered. Outlets and wall switches are left exposed.

Fig. 14-33. Kitchen cabinets hanging and attached to the completed soffit and wall.

Fig. 14-34. Finished particleboard floor before carpet is installed over it.

Fig. 14-35. Partially completed exterior showing the difference that the facing material can make over the exposed exterior wall sheathing.

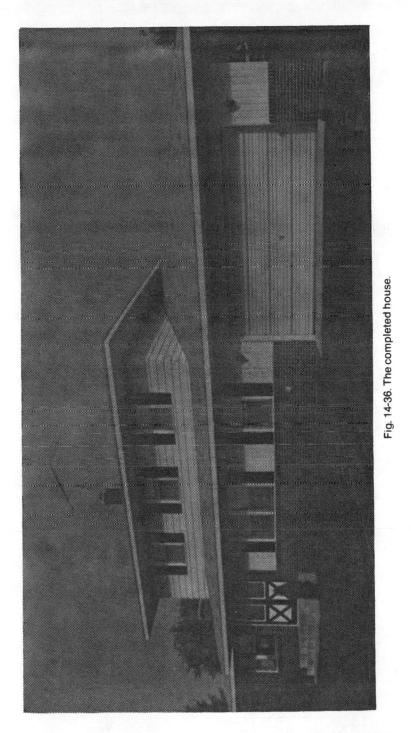

Fig. 14-36. The completed house.

Appendix

HANDY REFERENCES

Every handyman and homeowner faces the chore of buying lumber and associated hardware. This is particularly confusing because actual sizes (after the wood has been dried and cut) do not correspond to advertised sizes. Table A1 shows this difference.

Nails and screws can also be a problem. Table A2 shows the various sizes of nails, available in both flat and finishing heads. Figure A1 shows types of screws used for wood and metal. Also in Fig. A1 is the nomenclature used in sizing screws. Figure A2 demonstrates the proper method of sinking a screw.

Finally, it is sometimes necessary to convert between metric and English measurements. Table A3 provides an easy means of conversion for both length and area.

Table A1. Nominal Sizes and Standard Sizes of Lumber

Nominal size (in.)	American standard (in.)
1 x 3	25/32 x 2 5/8
1 x 4	25/32 x 3 5/8
1 x 6	25/32 x 5 5/8
1 x 8	25/32 x 7 1/2
1 x 10	25/32 x 9 1/2
1 x 12	25/32 x 11 1/2
2 x 4	1 5/8 x 3 5/8
2 x 6	1 5/8 x 5 5/8
2 x 8	1 5/8 x 7 1/2
2 x 10	1 5/8 x 9 1/2
2 x 12	1 5/8 x 11 1/2
3 x 8	2 5/8 x 7 1/2
3 x 10	2 5/8 x 9 1/2
3 x 12	2 5/8 x 11 1/2
4 x 12	3 5/8 x 11 1/2
4 x 16	3 5/8 x 15 1/2
6 x 12	5 1/2 x 11 1/2
6 x 16	5 1/2 x 15 1/2
6 x 18	5 1/2 x 17 1/2
8 x 16	7 1/2 x 15 1/2
8 x 20	7 1/2 x 19 1/2
8 x 24	7 1/2 x 23 1/2

Table A2. Sizes of Nails

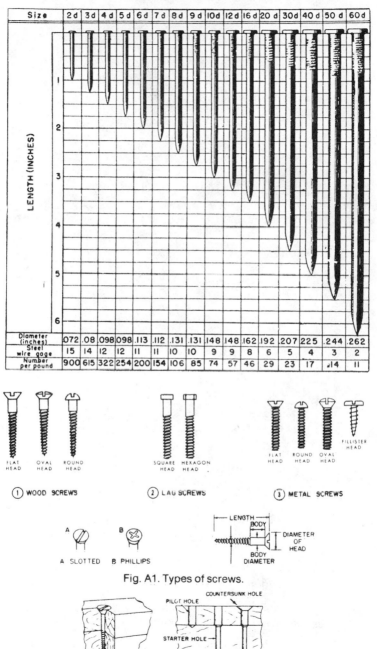

Size	2d	3d	4d	5d	6d	7d	8d	9d	10d	12d	16d	20d	30d	40d	50d	60d
Diameter (inches)	.072	.08	.098	.098	.113	.112	.131	.131	.148	.148	.162	.192	.207	.225	.244	.262
Steel wire gage	15	14	12	12	11	11	10	10	9	9	8	6	5	4	3	2
Number per pound	900	615	322	254	200	154	106	85	74	57	46	29	23	17	.14	11

FLAT HEAD OVAL HEAD ROUND HEAD

(1) WOOD SCREWS

SQUARE HEAD HEXAGON HEAD

(2) LAG SCREWS

FLAT HEAD ROUND HEAD OVAL HEAD FILLISTER HEAD

(3) METAL SCREWS

A SLOTTED   B PHILLIPS

LENGTH — BODY
DIAMETER OF HEAD
BODY DIAMETER

Fig. A1. Types of screws.

COUNTERSUNK HOLE
PILOT HOLE
STARTER HOLE

Fig. A2. Sinking a screw properly.

Table A3. Conversion Tables

Length

Metric to English

1 millimeter (mm)	=	0.04 inch (0.03937 inch)
1 centimeter (cm)	=	0.3937 inch
1 meter (m)	=	3.281 feet
1 m	=	1.094 yards
1 kilometer (km)	=	0.621 statute mile
1 km	=	0.5396 nautical mile

English to Metric

1 yard	=	91.44 centimeters (cm)
1 foot	=	30.48 cm
1 inch	=	2.54 cm
$7/8$ inch	=	2.22 cm (22.22 millimeters (mm))
$3/4$ inch	=	1.90 cm (19.05 mm)
$5/8$ inch	=	1.59 cm (15.88 mm)
$1/2$ inch	=	1.27 cm (12.70 mm)
$3/8$ inch	=	0.98 cm (9.84 mm)
$1/4$ inch	=	0.64 cm (6.35 mm)
$1/8$ inch	=	0.32 cm (3.18 mm)

Area

Metric to English

1 sq centimeter	=	0.155 sq inch
1 sq meter	=	10.76 sq ft
1 sq meter	=	1.196 sq yards
1 hectare	=	2.47 acres
1 sq kilometer	=	0.386 sq miles

English to Metric

1 sq inch	=	6.45 sq centimeters
1 sq foot	=	0.0929 sq meter
1 sq yard	=	0.836 sq meter
1 acre	=	0.405 hectare
1 sq mile	=	2.59 sq kilometers

Index

194